How the World Flows

How the World Flows

*Microfluidics from Raindrops
to Covid Tests*

ALBERT FOLCH

OXFORD
UNIVERSITY PRESS

OXFORD
UNIVERSITY PRESS

Oxford University Press is a department of the University of Oxford.
It furthers the University's objective of excellence in research, scholarship,
and education by publishing worldwide. Oxford is a registered trade mark of
Oxford University Press in the UK and in certain other countries.

Published in the United States of America by Oxford University Press
198 Madison Avenue, New York, NY 10016, United States of America.

CIP data is on file at the Library of Congress

ISBN 9780197772829
DOI: 10.1093/9780197772850.001.0001

Printed by Sheridan Books, Inc., United States of America

MIX
Paper | Supporting
responsible forestry
FSC® C008955
FSC
www.fsc.org

To Lisa,

For our laminar lifestream,

And to Jordi and Tàlia,

Who got mixed in it.

Contents

PART III. GOING WITH THE FLOW

Acknowledgments

This book has been a family project. It was my wife who insisted I write another book on microfluidics. The biologist in Lisa insisted that this one be for the general public and cover not just the academic microfluidic chips but also the mesmerizing microfluidics one finds in Nature and in a variety of contraptions and devices that have not been traditionally called "microfluidic." (I hope to change that.) I confide in her on many things—she has an MD/PhD in neurobiology and we have been working together in the lab for ten years—and reading (she reads a lot, I'm the writer) is her thing. I ditched several versions in our first discussions, so you will read the stories she pruned. On the other hand, her encouragement when the stories were "working" meant a lot. My father, who was a renowned editor in Barcelona and used to be my most trusted source in these matters, used to tell us: "Lisa has great taste in books." He also acted on this judgment: He published the translation of many books to Catalan after Lisa recommended them to him, most of which sold very well. Thus, every time she frowned at one of my story ideas was also a sweet reminder of my now-gone dad—and that I should keep trying. She also spent many hours with me restructuring the text, moving things around, and copyediting from beginning to end. As an author, it has been a luxury to enjoy the free consultation, literally at arm's length, of an editor-friend who also knows so much about my field and the life sciences. During one of these discussions overheard by my son Jordi, a budding chemical engineer, he suggested the title and we all thought it fit like a glove. And for the cover, I got help from my artistic daughter Tàlia.

Another form of encouragement, albeit indirectly, came from my parents. I had always wanted to explain to them why I love microfluidics so much without going too deep into the science. Science was not their thing, although they were both extraordinary thinkers. During the 1960s, they joined the pro-democracy movement against Franco's dictatorship, for which they were imprisoned a few times each briefly. For their respective careers, they both independently received the St. George Medal from the Government of Catalonia—the equivalent, if there is any, of the Presidential Medal of Honor in the United States. As microfluidics is rarely in the press, I thought I should explain my beautiful field to them. Unfortunately, my dad is no longer with us to help with the editing. Still, my mom's expertise as a sinologist and endless energy came in handy for the references to China in Chapter 11. She edited the text and every footnote to ensure that even an expert sinologist would approve them. She didn't get a St. George Medal for nothing.

I want to thank so many more people. First, I apologize to most of my microfluidic colleagues that this book only covers a tiny portion of the vast research and developments on microfluidic chips. Thus, it only mentions a few of them—those I could fit with the stories. I am very grateful to the many scientists and microfluidic colleagues I interviewed or contacted to get stories and/or associated images for the book: Cori Bargmann, David Beebe, Albert van der Berg, Todd Clark Brelje, Cullen Buie, Nikos Chronis, Dan Cohen, Andrew deMello, Guillaume Durey, Tal Dvir, Arjan Fraters, David Furness, Wei Gao, Hans Gardeniers, Jim Grotberg, Daniel Haber, Dan Huh, Don Ingber, David Juncker, Maria King, Imants Lauks, Thomas Laurell, Jennifer Lewis, Gang Li, Hang Lu, Andres Martinez, Richard Mathies, Ivar Meyvantsson, Avanish Mishra, Melur (Ram) K. Ramasubramanian, John Rogers, Houda Shafique, Shannon Stott, Massimo Temporelli, Mehmet Toner, Mark Wener, Dave Weitz, and Adam Woolley. I am particularly indebted to three amazing artists who use fluids in motion as their medium—Linden Gledhill, Frans Vandemaele, and Rose-Marie Lescure—for sharing their stories and photographs with me without asking anything in return but a simple acknowledgment line. I am also extremely thankful to Sara Stearns and the Chemical and Biological Systems Society for letting me use (again!) an extensive collection of beautiful photographs of historical microfluidic devices. Niall McDonald (Okaar Photography) took them for an exhibit at the MicroTAS'2016 conference in Dublin. Ellie Rose, the founder of Cocoa Press, a chocolate 3D printer manufacturer, provided two stunning pictures of chocolate 3D printing obtained by Kim Chase. My dear friend Patty Geldenblott, who read and professionally edited the stories from beginning to end, was instrumental in making them more readable and self-consistent.

There is a lot of borrowed material in this book. The book contains 124 figures comprising approximately 160 images. Behind each image that was not publicly available was an email request that, in most cases, prompted an enthusiastic answer, followed by a few more emails to finalize the picture I needed. Several individuals helped make my task much less daunting. Our close friend and oceanographer Vera Trainer taught me the difference between zooplankton and phytoplankton with her usual enthusiasm when it comes to the underwater microcosmos and pointed me to the beautiful image of Figure 15.5. I am grateful that there are vast communities of people online that believe in sharing materials freely (e.g., on Wikipedia, Wikimedia, Pxhere.com, Unsplash.com, Flickr, or Pixabay.com) and make them available even for commercial purposes such as this book. Without this community of generous, mostly anonymous contributors, I would not have been able to illustrate these pages with so many beautiful images. When the photographer's name was posted, I placed their name on the figure caption as a small thank-you note, even if the license did not require it. To complement the book, I recommend visiting the YouTube Channel "The Lutetium Project," where one can find superb introductory videos to the science and technology of microfluidics. I'd also like to collectively thank my whole group for producing a wealth of gorgeous photos that inspired me to start BAIT in 2007. BAIT—short for Bringing Art Into Technology—is an outreach art program that entices scientists and nonscientists alike to inquire about our scientific projects using striking imagery and movies. I remain convinced that the BAIT images will outlast the science that originated them. Typically, my students take the photos, and some of these end up being used for exhibitions in public halls, figures in books such as this one, and brochures. Beautiful art can be an effective "bait" to catch people's attention to science. With that in mind, almost none of the images in the book contain scale bars, and I have simplified any schematics to depart from the textbook-figure style.

The last acknowledgment goes to my postdoc mentor and friend, Mehmet Toner—for his contagious vision, relentless enthusiasm, and unmatched integrity. More than a quarter century ago, Mehmet patiently taught me to think about the applications of microengineering in biomedicine and, in particular, about microfluidics—why small things tend to work better *mechanically*, why one should *use* them in the first place, and when *not*. An added benefit of his company is that he always turns to his sense of humor when the conversation has gone too serious—so we laugh a lot. And he always encouraged—and continues to encourage—me to write. Now, add every reader's debt to Mehmet—many ideas he planted in my head are peppered throughout this book—and you'll get a measure of my immense debt—and gratitude—to him.

Introduction

What Is "Microfluidics"?

If my mind follows the wandering path of stars
Then my feet no longer rest on Earth.

—Ptolemy (ca. second century AD), in *The Almagest*

I suspect that I entered the study of physics with the Ptolemaic hope, like many in my generation, that I would become elevated by studying the large phenomena of the universe. But Science takes many detours, and after a trip to the Lawrence Berkeley Labs, overlooking the Bay Area, during my PhD, I found myself peering down into microscopes instead. Yet my journey through the world of the small became a broadening, illuminating experience. And what brought the light, surprisingly, was water.

We have an intimate relationship with—one could say a debt to—water, not the least because approximately three-fourths of our brain and muscles are made of this unique liquid that, coincidentally, covers almost three-fourths of Earth's surface. Fortunately, our planet is splashed with oceans, lakes, ponds, rivers, and waterfalls. It's not just that they make beautiful vacation spots and allow for cooling down in the summer. Without them, there would not be life.

Kneel on the shore of a marsh or a lake, and, looking at close range, you will discover a whole community of organisms that live on its surface. You will find tiny creatures clinging upside down. As if endowed by magic, some insects can glide over water just as an ice skater slides over a frozen pond. Yet, when you try to clench your fist around a fluid volume, unlike a solid body, you cannot break it. It always slips through your fingers, like air—yet we cannot throw a stone while submerged inside water like we do in open air. Why are liquids so mesmerizing and why do they behave so oddly when we look at them with a magnifying lens?

These unique properties of liquids have fascinated inquisitive minds for centuries. If you could inspect every raindrop that falls from the sky to fill a puddle, you would find that each contains a speck of dust and occasionally a bacterium or a virus, likely lifted by the wind from a faraway swamp. And water drops can

Figure I.1 Water droplets can act as tiny lenses.

Reproduced from Elbetiko (2018). Droplets Reflection. Pixabay. https://pixabay.com/photos/droplets-reflection-flower-plant-3263600/. Public domain license.

bend light rays like a lens (Figure I.1). Each droplet is a miniature ecosystem; its surface is a gateway to light, essential gases, and nutrients. And each cell in our body, including every cell in every plant, is a tiny volume of fluid surrounded by a flimsy membrane that serves to confine the hundreds of biochemical reactions that we call Life.

The chapters in this book strive to act like a microscope that pulls the reader into this barely noticeable, Lilliputian world of fluids—the *microfluidic* world. Just like microelectronic circuits are too small for you to see and are hidden under the screen of your smartphone, most microfluidic phenomena happen in tiny spaces such as a cell; a dewdrop; your blood capillaries; and minuscular crevasses between paper fibers, grains of sand, and inside manufactured devices such as glucometers and inkjet printers.

I have organized the eighteen chapters of the book into three parts that present different microfluidic systems of increasing complexity: the simpler droplets (Part I), wicking into small conduits called *capillaries* (Part II), and the more complex microfluidic systems that involve continuous flow (Part III). Although, arguably, cells are also microfluidic entities, the (occasional) fluid transport between cells occurs on a much smaller scale, so I left it for some other time. Each of the six chapters in each part tells a story related to microfluidics.

In most stories, the characters are everyday people (a sailor, a housewife, two farmers, a chef, a screenwriter, a rapper, and a doctor). In two of the stories, however, the protagonists are (*very*) famous scientists because I wanted to give the reader a tangible grasp of how microfluidics has been occupying the time of great minds for a long time: it's not a minor subject. Although independent, the stories also present key scientific and engineering concepts that are best understood if read in the presented order. Chapter 6, which belongs to Part I (on droplets) and deals with droplets' applications to paint and printing, also covers 3D printing despite not being a droplet technology for the most part. This anomaly is justified because all droplet generators and 3D printers use microfabricated nozzles; also, the transition from droplets (paint sprays and inkjet printers) to a continuous filament (3D printers) is due to a difference in the fluid's viscosity—which makes for a good lesson for the reader.

Perhaps as important as these three visible parts are the three subjacent themes sprinkled throughout each chapter. First are the microfluidic systems made by Nature through evolution and natural selection, from raindrops to the sap of plants all the way to the unmatched sophistication of the gills and the kidneys. Second are the rudimentary microfluidic technologies devised by humans, from candle wicks (ages ago) to paint sprays (during the Industrial Revolution). And third, engineers have manufactured many *microfluidic chip* devices using miniaturization technology—from the nozzles of inkjet printers to pregnancy tests and more. All three types of microfluidic systems have in common that despite their microscopic size, they have had a colossal impact on our planet and our lives. In some capacity or another, they all make the world flow.

The list of scientists who paid attention to microfluidic wonders in the past few centuries—mainly the study of wicking or *capillarity*—is long and illustrious. However, these researchers—primarily chemists and physicists—were limited by the tools at their disposal until recently. A new breed of engineers (we now call them "microfluidic engineers") learned to repurpose microelectronics technology to manipulate fluids. These first microfluidic devices, like the first transistor, were not very sophisticated by today's standards. In 1961, C. C. Mattax and J. R. Kyte, two oil engineers of the Jersey Production Research Company, a subsidiary of Standard Oil (the company that became Exxon Mobil), wanted to understand the oil extraction process better. Because they could not place a camera down the shaft, they built a transparent micromodel of water-flooded dirt. The "dirt" was a set of tiny features and channels etched in glass—the first known microfluidic chip.[1] Their innovation went unnoticed outside the field

---

[1] Mattax, C. C., & Kyte, J. R. Ever see a water flood? *Oil Gas J.* 59: 115–128 (1961).

of oil engineering, and microfluidic chips were invented and reinvented several times over three decades.[2-4]

Microfluidics has become a full-grown engineering subdiscipline which deals with devices that use fluids confined to less than 1 millimeter in at least one dimension. The liquid is usually confined in channels, tiny pores, or droplets. Much like in microelectronics, size is critical in microfluidics. As the components get smaller, devices rely on the strange properties of liquids at small scales, are cheaper to fabricate, and operate faster and more efficiently. The devices can reduce reagent consumption, control the mixing of chemical and biological substances, speed up the manipulation of single cells and other particles, and automate multiple experiments in one chip with microvalves while facilitating imaging and various measurements—the so-called lab-on-a-chip. The microfluidics revolution has been silently piggybacking on its electronic counterpart without making any headlines.

More than sixty years after Mattax and Kyte made their first chip, the deployment of microfluidic chips in society is vast, in contrast to their tiny size—akin to the wide distribution of minuscule microelectronic chips that form the brains of smartphones, televisions, computers, car consoles, microwave ovens and fridges, Wi-Fi routers, you name it. Yet microelectronics seems to catch people's eyes differently compared to microfluidics. Microelectronics is a popular subject because almost every adult human watches television and walks around with a smartphone, uses it to organize their life, and cares about their device getting wet or running out of battery. On the news, we are constantly reminded of the mightiness that microelectronic chips bestow and the challenges they pose. They fuel artificial intelligence. They heat up, so data storage centers have to refrigerate them, which consumes large amounts of electricity. A single company in Taiwan manufactures 90 percent of them, triggering geopolitical frictions and causing chip supply shortages during the pandemic. The chips contain tiny amounts of metals of challenging and controversial mining. And so on.

But pick an average person tapping on a smartphone, and, likely, they have not even heard the term *microfluidics*. Perhaps it's not surprising because the word "microfluidics" was not coined in the scientific literature until 1992.[5] My goal in writing this book has been to add this uncommon word to the common parlance. Still, although microfluidics has impacted people just as much as microelectronics, it is not a buzzword yet. Did you know that the COVID-19

[2] Little, W. A. Microminiature refrigeration. *Rev. Sci. Instrum.* 55: 661–680 (1984).

[3] Wu, P., & Little, W. A. Measurement of the heat transfer characteristics of gas flow in fine channel heat exchangers used for microminiature refrigerators. *Cryogenics* 24: 415–420 (1984).

[4] Walter, J., Kern-Veits, B., Huf, J., Stolze, B., & Bonhoeffer, F. Recognition of position-specific properties of tectal cell membranes by retinal axons in vitro. *Development* 101: 685–696 (1987).

[5] Bloomstein, T. M., & Ehrlich, D. J. Laser-chemical three-dimensional writing for micro-electromechanics and application to standard-cell microfluidics. *J. Vac. Sci. Technol.* B 10: 2671 (1992).

tests we monitored our snot with during the pandemic, the pregnancy tests women use to predict if they are expecting a baby, the glucometer that diabetic people rely on daily to stay alive, and the printheads of inkjet printers and 3D printers are *all* microfluidic devices? Developed in the wake of microelectronics, these microfluidic devices have impacted the lives of billions of humans. Many other microfluidic chips are being developed for personalized medicine and drug screening, among other applications.

Yet, this book did not have space to cover the many types of microfluidic chips, so this is not a book for students to learn microfluidic engineering. Our field is already very well covered by several textbooks and hundreds of reviews, including my previous book that delved deeper into microfluidic chip history (*Hidden in Plain Sight: The History, Science, and Engineering of Microfluidic Technology*; MIT Press, 2022). I apologize to my many microfluidic colleagues who will justifiably feel they deserved a mention here. They are doing a lot of fascinating research, and there is much work ahead. In the book you have in your hands, I was aiming for a broader perspective that would do justice to the foundations of our field: Many microfluidic devices precede microelectronics and microfluidic chips by a long, long time. Microelectronic chips are relatively recent—some engineers who created the first ones are still alive—and so are microfluidic engineers. The candle wick, the paintbrush, the ballpoint pen, the carburetor, and the kidney dialysis machine are examples of "pre-chip" microfluidic devices. The space devoted in this book to these venerable microfluidic devices necessarily reduced the space available for the modern microfluidic chips developed in the past three decades. One could easily argue that the ballpoint pen and the fuel injector have impacted humanity in a way that far surpasses microelectronics.

And, as an exercise in humility, we need to acknowledge that Nature has been developing microfluidics since the beginning of time. First and foremost, there are the natural raindrops, without which there would not be rainbows, agriculture, or life on our planet. Plants use capillarity to transport nutrients dissolved in sap from their roots to the highest branches. Water evaporation occurs at tiny pores in the leaves connected to the sap channels, creating a microfluidic suction effect that causes the sap to ascend for as long as the plant is alive—which can be more than a thousand years for some trees. And we, like plants, are also microfluidic: Tiny blood vessels distribute nutrients and oxygen to every cell of our bodies. We are still struggling to understand how Nature came up with these designs. Still, we know this much: Our planet would look very different, and we certainly would not be here, without microfluidics.

This book invites you, the reader, to a journey of discovery through the transformative power of microfluidics. If microfluidics becomes a new buzzword for you, I will have done my job. I hope these pages can serve as an illuminating, magnifying lens so that you can also marvel at these minuscule wonders that surround us everywhere.

PART I

A WORLD OF DROPLETS

But words are things, and a small drop of ink,
Falling like dew, upon a thought, produces
That which makes thousands, perhaps millions, think.

—Lord Byron (1788–1824)

Tiny water droplets surround us everywhere. Despite being among the smallest and simplest microfluidic systems, they have an enormous bearing on our lives. A cubic meter of cloud, fog, or mist contains several hundred million droplets, each separated from its nearest neighbor by about 1 millimeter and only 10 or 20 micrometers in diameter.[1] Also known as a *micron*, a micrometer is one-thousandth of a millimeter. This size makes water droplets inside clouds so light that they are lifted by rising warm air, and wind carries them away. Each droplet is perfectly transparent, but when the sunlight hits the cloud, the droplets reflect the sun's rays in all directions, giving the cloud its cotton candy appearance.[a] Droplets suspended on winds carry dirt, bacteria, and fungi across oceans, covering entire continents with soil, life, and diseases when they drop their cargo as raindrops, mist, or dewdrops. And let's not forget that our bodies generate sweat drops, tears, and the controversial spitting droplets of our breath and cough, against which we were instructed to wear masks during the COVID-19 pandemic.

Many of the droplets that exist around us are manufactured by humans. The sauces we eat or the moisturizers and shampoos we spread on our skin are suspensions of water droplets in oil called emulsions. The "jets" of inkjet printers and some 3D printers are tiny streams of bullet-like ink drops precisely shooting

---

[a] Droplets inside high-altitude clouds can freeze into ice crystals and coexist with droplets "super-cooled" below 0°C that are too small to freeze. Freezing does not change their weight, so that does not cause them to precipitate. Internal cloud air drafts are responsible for collisions between ice or snow crystals and droplets. These collisions cause the drops to grow and fall in the form of rain, snow, or hail, depending on the temperature.[1]

[1] National Weather Service. *Cloud development.* https://www.weather.gov/source/zhu/ZHU_Training_Page/clouds/cloud_development/clouds.htm (n.d.).

through the air used to mark, paint, or build many everyday objects. All spray devices—a nebulizer for asthma patients, a window cleaner, an airbrush paint tool—project broad bouquets of droplets into the air. If you could peer into all the combustion engines running since their invention almost a century and a half ago, you would find that, in every single model, the fuel was mixed with air before ignition by generating a spray. We have been depending on microscopic droplets for eating, healing, painting, printing, and locomotion—most of the time without realizing it.

Chemists have designed microfluidic devices that, mimicking the physical properties of raindrops, subdivide a sample such as a speck of blood into millions of droplets and analyze them at dizzying speeds. Each droplet is essentially a tiny chemical laboratory, and these microfluidic devices allow for massively parallel genetic analysis and studying the evolution of biomolecules, among other things. It's a world of droplets out there.

1

The Liquid Pearls of Mother Nature

The Microfluidics of Rainbows, Clouds, and Other Aerosol Phenomena—and How Droplets Are Affected by Humidity, Surfaces, and Diffusion

In 1665, rainbows inspired a well-off, half-orphan 22-year-old student at Trinity College in Cambridge, England, to study the nature of light. He looked at raindrops for an answer. Despite their brief existence and tiny size, these microfluidic droplets in suspension put on magnificently large and colorful displays when illuminated by sun rays: rainbows (Figure 1.1).

The young student had come from a family of wealthy Lincolnshire farmers but had not had a happy childhood. He had never known his father, who died months before he was born, and had been living with his grandmother in the manor house in Woolsthorpe, near Grantham, because his mother had to remarry. He had not shown much promise in the Free Grammar School in Grantham, but an uncle decided he should go to the university. In the summer of 1665, he had just received his undergraduate degree and was looking forward to pursuing his master's. Then the Great Plague struck.

The university closed, and he had to return to his Woolsthorpe home while the pandemic raged for well over a year. The forced isolation gave the young man plenty of time to think and do experiments on his own. He pondered the nature of things that he saw around him, so he started with light and rainbows. During this time, however, he would not limit himself to these two important subjects—he also invented calculus and came up with the law of gravitation. He would become one of history's most renowned and influential scientists, but he was just getting started. His name was Isaac Newton.

Newton's passionate curiosity attracted him to find ways to manipulate light. He devised a revolutionary experiment using a simple object that had been

Figure 1.1 Rainbow formation after rain.
Image source: Pxhere.com. https://pxhere.com/en/photo/939486.

greatly admired[a] but never so rigorously examined before: the *prism*, a glass crystal separating white light into a rainbow of colors (Figure 1.2). In 1665, according to his notebook,[1] Newton decided to conduct research with a glass prism that he had obtained at the Stourbridge Fair, an annual market held in Cambridge. He darkened his study and let a narrow beam of sunlight pass through "a round hole about one third part of an inch broad"[2] (8.5 mm in diameter) in the shutters. Next, he placed the prism in the sunbeam. As expected, he then saw a rainbow projected on the opposite wall, or in his own words, "a coloured image of the Sun."[2] He verified that a second prism placed at the wall was able to project back white light; in other words, the splitting of light was reversible. From this homemade rainbow, he was the first to correctly infer that white light comprises seven colors: red, orange, yellow, green, blue, indigo, and violet. We now know that these colors compose the visible part of a broader

---

[a] The ability of crystals to project colors had fascinated Roman writers such as Seneca and Pliny the Elder.[3] When Marco Polo arrived in China around 1271, after his three-and-a-half-year-long journey from Venice, he brought one such crystal on the Pope's behalf as a gift for emperor Kublai Khan.[4] Four centuries later, in Newton's time, prisms had become everyday decorative objects in the West.

[1] Hall, A. R. Sir Isaac Newton's note-book, 1661–65. *Cambridge Hist. J.* 9: 239–250 (1948).

[2] Newton, I. *Opticks: Or, a Treatise of the Reflexions, Refractions, Inflexions and Colours of Light. Also Two Treatises of the Species and Magnitude of Curvilinear Figures.* Sam. Smith and Benj. Walford (printers to the Royal Society), 1704.

[3] Sparavigna, A. C. The play of colours of prisms: A short history of prisms from Lucius Anneus Seneca to George Ravenscroft. *arXiv* (2012).

[4] Polo, M. *The Travels.* Penguin (1958).

Figure 1.2 Separation of white light into a rainbow by a glass prism.

Image source: User Braxton Apana on Unsplash.com. https://unsplash.com/photos/ person-holding-white-box-with-rainbow-light-VuNNRTFdrME.

electromagnetic spectrum that covers other forms of light waves we cannot see, such as infrared, radio, gamma, microwaves, and ultraviolet radiation.

Newton's findings shattered many beliefs. During Antiquity and the Middle Ages, the rainbow's wondrous colors, grandiose size, ephemeral existence, and connections to sunlight and rain stood as irrefutable proof of the majestic and inscrutable power of the gods. The inability of humans to explain it gave rise to countless myths. The Greeks (and later the Romans) believed that the rainbow was a path made by the messenger Iris from Heaven to Earth. In Irish folklore, the leprechaun, a solitary and mischievous little fairy—a bearded shoemaker dressed in green—hides a pot of gold at the end of the rainbow. The point is that there are as many rainbow myths as cultures.[b] As amusing as these myths may sound in our present era based on rational experimentation, they did play a role in providing some explanation for why that luminous arch appeared up in the sky.

Light itself was very puzzling for a long time. Lens-shaped objects with light-bending properties have been found in archeological sites in the eastern Mediterranean since the Bronze Age.[5] Euclid—the Greek mathematician considered the "father of geometry"—mistakenly thought that rays were emitted from the eyes and bounced back from objects.[6] This theory went undisputed until the eleventh century AD when Persian physicist Ibn al-Haytham—one of the most brilliant scientists of medieval Islam—experimented with bulls' eyes and correctly concluded that light became bent or *refracted* by a lens inside the eye.[7] Lenses achieve their light-bending effect because the light changes its speed when it travels from one material to another, and in doing so, the ray also changes its direction proportionally to its change in speed. Because of all the changes undertaken by light upon traversing materials, everyone until Newton erroneously thought that light's color, too, was a mutable property. Newton, by contrast, correctly surmised that color had to be an immutable property of light when he found that the rainbow projected with the prism on the wall of his darkened living room could be reversed back to white light with a second prism.

After the prism experiment, Newton rightly inferred that the colors of rainbows were attributable to raindrops suspended in the air just after a shower. The raindrops acted as tiny miniature liquid lenses due to their deliciously round shape and pristinely smooth surfaces (see Figure 1.1). The sun's rays hit the curved surface of the drops at an angle, causing the bending of light (Ibn al-Haytham's *refraction*) as each light ray penetrates the drop and reflects inside the drop like a mirror. Newton did not know that light was an electromagnetic wave—a concept proposed by James Clerk Maxwell in 1865—or that it traveled at different speeds through different materials depending on its energy (i.e., color).[c] We now know that the Sun's white light is a

[b] Examples abound. In the Old Testament story of Noah, after God is done flooding Earth to wash away its corruption, He places a rainbow in the sky as a sign that He would never exact such a devastating punishment again. In Peru, a pre-Incan superstition encourages people to close their mouths at the sight of the rainbow to avoid getting sick.

[c] Newton erroneously attempted to explain refraction by proposing that light was composed of particles. Although refraction can only be explained when considering light as an electromagnetic wave, Newton's corpuscular theory of light was partially rescued in the twentieth century. To explain certain phenomena, such as the photoelectric effect, the new theory of quantum mechanics proposed that light has a dual wave–particle nature.

[5] Plantzos, D. Crystals and lenses in the Graeco-Roman world. *Am. J. Archaeol.* 101: 451–464 (1997).

[6] Woodford, C. *Scientists Who Changed History*. DK Publishing (2019).

[7] Omar, S. B. *Ibn-Al-Haytham's Optics: A Study of the Origins of Experimental Science*. Bibliotheca Islamica (1977).

mixture of waves, each representing a different color. As all the mixed sun-beams or light waves enter a droplet in the air, they slow down and thus refract (bend) toward the inside of the droplet at an angle that depends on the speed at which they propagate in water. The red and orange color waves are the ones that slow down and refract the least. In contrast, the blue and purple color waves slow down and refract the most—which accounts for the fan-like separation in a prism and the projection of colors in the rainbow.[d]

If you are dazzled by the magic of the rainbow, the conclusion you would draw from Newton's experiments might be rather gloomy: Neither the rain nor the Sun is strictly necessary for the rainbow to appear—you can create it yourself by spraying a mist of water with your back against a white floodlight. Newton saw in the serene rainbows a microfluidic dance with light: As you look at any given spot of a rainbow, hundreds of droplets are settling down or drifting away in the breeze, each droplet taking the place of the preceding one to deflect the sunrays the same way towards your retina. Although New-ton's corpuscular refraction theory of light was incorrect, his effort at reducing the enigmatic rainbow to its mechanism of droplets was spot on and illu-minated the dark skies of superstition, ultimately ending the mythical beliefs regarding the origin and end of the rainbow. As two different observers will see the light emitted from different groups of droplets, they will never see the same rainbow. Similarly, as you move a few steps toward the end of the rainbow, a different set of droplets projects a new rainbow a bit farther away. The end of the rainbow is unreachable. The droplets broke the leprechaun spell.

Without Iris, the messenger, flying through the clouds, Newton's successors faced the challenge of explaining how these rainbow-producing droplets form in mid-air—not a small task. Since Newton, we have learned that drops grow by *condensation*, the process by which a gas converts to its liquid form, and that we do not need to wait for a rainbow to observe it. The condensation occurring on a cold morning receives the poetic name of *dew* (Figure 1.3), another beautiful

---

[d] Under very clear conditions, you can even observe two rainbows in the sky, one above the other (see Figure 1.1). The brighter one, lower in the sky, is called the primary rainbow. The one higher up in the sky, called the secondary rainbow, is twenty times fainter. The sunlight produces two rainbows because the beams reflect off inside the droplet and bend again as they exit each drop toward you. For the primary rainbow, the exit angle is between 40 degrees (for violet light) and 42 degrees (for red light) with respect to the entrance angle. For the secondary rainbow, the exit angle is inverted in order, between 50 degrees (for red light) and 52 degrees (for violet light), because the sunbeams have undergone an additional reflection inside the drops. For all the angles in between, there is no refrac-tion (no colors), but there are still many sunbeams bouncing off the internal walls of the droplets, and in each bounce a small amount of light is lost. For this reason, when the primary and the secondary rainbows appear together in the sky, an enigmatic-looking—subtle—dark band separates the two rainbows.[8]

[8] Haußmann, A. Rainbows in nature: Recent advances in observation and theory. *Eur. J. Phys.* 37: 063001 (2016).

Figure 1.3 Dewdrops on a grass leaf. Note how the water surfaces cannot spread on the water-repellent leaf and bead up.

Image source: Ju Irun on Pixabay. https://pixabay.com/photos/leaf-droplets-reflection-grass-2986837/ (CC0).

display of Nature produced by drops, like the rainbow, also evoked by many writers.[e]

Surprisingly, condensation cannot occur on a pristine surface—it always starts around minute features of a surface, such as a crack or a dust speck. Around these "seeds," microscopic droplets gradually grow into visible drops by further condensation. And even then, surfaces continue to influence the life of drops beyond condensation. To control evaporation and hydration, cells on the surface of leaves, grass, and fruit secrete a protective film of wax. This film is *hydrophobic*, or water-repellent, causing water to bead up rather than spread.

Dewdrops and the raindrops on your windshield may look very still to you, but a molecular storm is raging inside. At room temperature, water molecules move around at the dizzying average speed of more than 2,000 kilometers per hour—more than twice as fast as an airplane. However, the trip is relatively short because each water molecule collides a few trillion times per second with other water molecules. In a fraction of a trillionth of a second, they travel about one length of their molecular body length (a quarter of a billionth of a meter) before colliding against another molecule. It's a packed frenzy in there.

The sum of all the energy required to move these molecules (including their vibration and rotation) is what we vulgarly know as *heat* and gives away the *temperature* of the fluid. With these thermal motions, molecules in solution naturally tend to spread far away from their initial position as they bounce with each other. The scientific name for this tendency is *diffusion*.

[e] A famous example is the second verse of Shakespeare's *Romeo and Juliet*, in which the poet describes Montague's sadness "with tears augmenting the fresh morning's dew."

A concept now well grasped by high schoolers, it was not until 1855 that young German physiologist Adolf Fick formulated the *laws of diffusion* that predict that the spread of a molecule's concentration in a liquid evolves *quadratically* rather than linearly with time. In other words, molecules dissolved in liquids—such as the red dye you can see in Figure 1.4—do not move at a constant speed like cars on a freeway usually do. Instead, with all the bumping between the water molecules, it takes the red dye molecules on average *four times as long to travel twice as far.* Some take shorter paths and some longer ones, but on average, this law is inescapable: It applies to any molecule dissolved in any solvent. In practice, molecules diffuse fast over short distances but very slowly over long ones, forming *concentration gradients* (see Figure 1.4).

All this molecular activity makes droplets very vulnerable to evaporation. A droplet exposed to air continuously exchanges water vapor molecules with its surroundings (always via diffusion). If the environment is hot and dry, the droplet will rapidly shrink by evaporation and disappear. On the other hand, inside a sauna, your newly formed droplets of sweat are in a very humid air saturated with water vapor molecules and lose as many water molecules as they gain and they don't seem to evaporate. Likewise, during the cold, still, and humid mornings that favor dewdrop formation, droplets are momentarily in equilibrium with their surroundings before the first rays of the sun start heating the air and the leaves. Every brief display of dewdrops is a delicate reminder of how beautifully balanced Nature is.

Leaves and grass provide a condensation surface for the morning dew, but how can water condense in mid-air into airborne droplets to form clouds and fog? In 1881, Scottish engineer John Aitken performed experiments to unveil that airborne droplets cannot grow in perfectly clean air. Then, how do cloud droplets form? Similar to how an oyster needs an impurity to grow a pearl, particles floating in the air can act as a "seed" for the growth of a droplet. In an article titled "On Dust, Fogs, and Clouds," Aitken wrote,

Dust is the germ of which fogs and clouds are the developed phenomena.. . . Our breath when it becomes visible on a cold morning, and every puff of steam as it escapes into the air, show the impure and dusty condition of our atmosphere.[9]

These airborne particles in the atmosphere that nucleate the growth of every drop in every cloud, mist, or fog, as discovered by Aitken, are now called *aerosols.*[10]

---

[9] Aitken, J. On dust, fogs, and clouds. *Nature* 23: 384–385, 195 & 196 (1881).

[10] Ramanathan, V., Crutzen, P. J., Kiehl, J. T., & Rosenfeld, D. Atmosphere: Aerosols, climate, and the hydrological cycle. *Science* 294: 2119–2124 (2001).

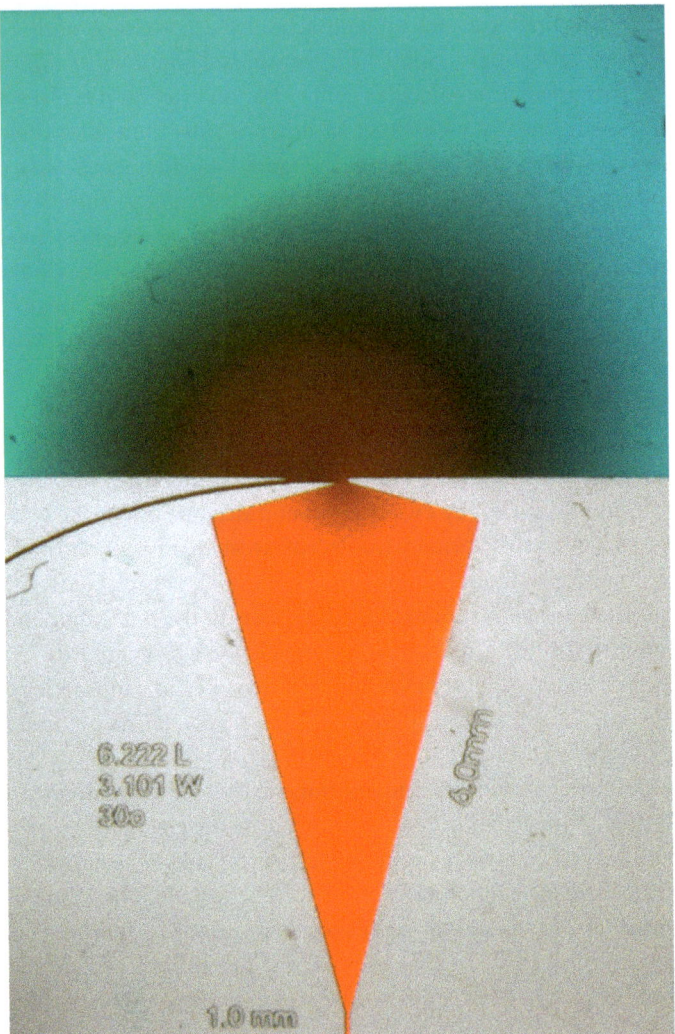

Figure 1.4 The diffusion of red dye in a microchannel filled with blue dye produces a gradient that evolves with time.
Image source: Tom Keenan and Albert Folch.

Even before humans introduced industrial pollution, the atmosphere naturally contained abundant suspended particles too small to be seen or to settle by gravity. That's why clouds have always existed on Earth—and in all known planets with an atmosphere. You are breathing aerosols in right now. These floating particles are typically less than 1 micrometer (a thousandth of a millimeter) across. Each year, seasonal winds lift tens of millions of tons of sand from the Sahara

Desert in North Africa and carry massive plumes of dust toward the Americas and over Europe. Massive wind systems such as the monsoons act as a periodic shuttle for aerosols between the oceans and the continents. Aerosols are *mostly* dust but not *only* dust. Ocean spray lifts salt crystals. Various sources generate ash particles—for example, forest fires, volcanic eruptions, and, more recently, those created by industrial fumes and car exhaust. Every single raindrop and snowflake has a dirt speck inside.

There are also *living* aerosols.[11] Somehow, a constellation of spores, pollen, fungi, algae, bacteria, and viruses gets lifted into the atmosphere along with the sea spray, dust, and soot. And every time animals cough, sneeze, yawn, or exhale, they project many organisms inside droplets. French chemist and biologist Louis Pasteur was the first to prove the abundant presence of microorganisms in the air in 1858. A recent study determined that between 20 and 70 percent of the particles in the atmosphere are of biological origin, with the number of cells ranging from 100 to 10,000 per cubic meter.[12] A scary example of this type of aerosol is the one we produce when talking, coughing, or sneezing (Figure 1.5). This mechanism is responsible for transmitting infections such as the flu or the COVID-19 virus, which had most humans confined to their houses and wearing masks for months. Very few people would have suspected, just a few years ago, that something as tiny as droplets could have such a massive impact on our lives.

Dr. Gavin Koh, a Singapore-born physician based in Cambridge, England, who specializes in infectious diseases, had suspected for quite some time that aerosols could be the transmitting vehicle of some bacterial infections. He is an expert in melioidosis,[f] a severe infectious disease caused by the bacterium *Burkholderia pseudomallei* present in the soil, decaying organic matter, and surface water in much of the tropics. The two most severe manifestations of melioidosis are pneumonia and blood poisoning, resulting in fever, difficulty breathing, and a collapse in blood pressure. Patients often need treatment in intensive care hooked up to life support, which is not always available in rural areas. Untreated, melioidosis has a mortality of more than 90 percent. Even with antibiotic treatment, mortality is 20–40 percent.

Although person-to-person transmission is extremely rare, the bacterium spreads to humans from the soil with startling ease. Researchers such as Koh tried to figure out the contagion mechanism to prevent the disease, but it had

[f] Melioidosis is endemic in tropical areas, especially in Southeast Asia and northern Australia, affecting 165,000 people and killing 89,000 individuals every year.[13]

[11] Després, V. R., Huffman, J. A., Burrows, S. M., Hoose, C., Safatov, A. S., Buryak, G., et al. Primary biological aerosol particles in the atmosphere: A review. *Tellus* 64: 15598 (2012).

[12] Dommergue, A., Amato, P., Tignat-Perrier, R., Magand, O., Thollot, A., Joly, M., et al. Methods to investigate the global atmospheric microbiome. *Front. Microbiol.* 10: 1–12 (2019).

[13] Joung, Y. S., & Buie, C. R. Aerosol generation by raindrop impact on soil. *Nat. Commun.* 6: 6083 (2015).

Figure 1.5 Photograph of a man mid-sneeze, revealing the plume of salivary droplets as they are expelled in a large cone-shaped array from the man's open mouth. The picture, taken in 2009, dramatically illustrates the reason people need to cover their mouth when coughing, or sneezing, in order to protect others from germ exposure.

Image source: James Gathany, Centers for Disease Control and Prevention. https://en.wikipedia. org/wiki/Respiratory_droplet#/media/File:Sneeze.JPG. Public image.

remained elusive. For many years, he and his colleagues had noticed outbursts of infections coinciding with heavy tropical rain storms. And in the Vietnam War, helicopter pilots were particularly susceptible to melioidosis. These observations suggested that the bacteria somehow became airborne—but how?

In January 2015, a research report by Massachusetts Institute of Technology professor and microfluidic engineer Cullen Buie caught Koh's keen eye across the Atlantic. Using high-speed video recording of raindrops impinging on a laboratory surface, Buie showed how an impacting drop can sprinkle aerosols up the air (Figure 1.6).[14] He did not know then that it could also have far-reaching

[14] Wiersinga, W. J., Virk, H. S., Torres, A. G., Currie, B. J., Peacock, S. J., Dance, D. A. B., & Limmathurotsakul, D. Melioidosis. *Nat. Rev. Dis. Prim.* 4: 17107 (2018).

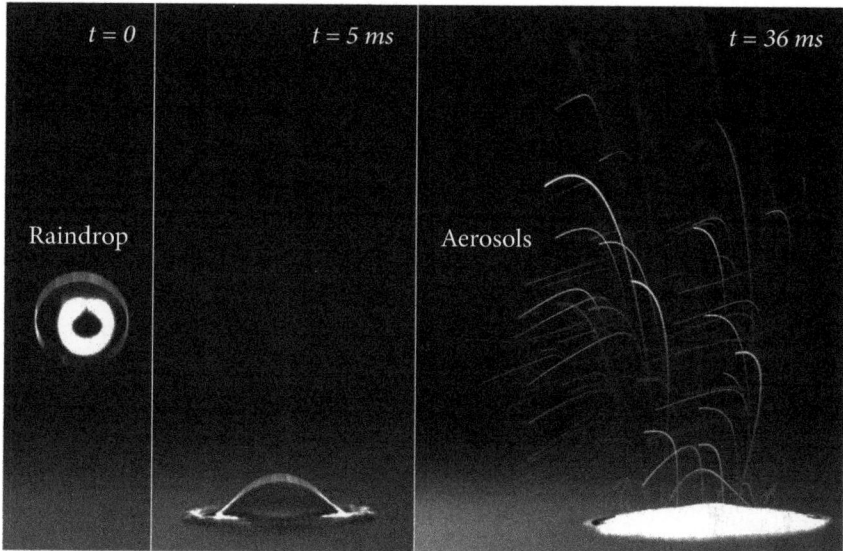

Figure 1.6 Sequence of three images showing the aerosols generated upon impact from a raindrop falling onto a laboratory surface. The elapsed time is indicated at the top of each image in thousandths of a second (abbreviated as "ms" or milliseconds). The drop in the air about to hit the surface in the first image is about 3 millimeters in diameter.

Images courtesy of Prof. Cullen Buie.

implications on human health. Koh immediately contacted Buie with a hypothetical question: Could this mechanism lift bacteria into the air so that they
could become inhaled by a person? Buie was intrigued. His student Young Soo
Joung repeated the experiments, with bacteria on the surface this time, which
confirmed one plausible mechanism for microbes taking microfluidic shuttle
trips into the sky: The impact of a raindrop can indeed eject soil bacteria up the
air inside the aerosols, where they can remain viable for at least 1 hour.[15] Similarly long "hang times" have been reported for virus-carrying cough or sneeze
aerosols.

In the air, when they are not already wrapped up in fluid from the start, these
mineral and biological particles become the seed for the condensation—or simply growth—of droplets around them, forming clouds of various shapes and
sizes. The clouds effectively become the home of droplet-encased particles until
they get redeposited elsewhere, perhaps on another continent or over a distant
sea or lake—or directly on the open wound or the inner linings of the lungs of an

[15] Joung, Y. S., Ge, Z., & Buie, C. R. Bioaerosol generation by raindrops on soil. *Nat. Commun.* 8:
14668 (2017).

organism—where there is ample supply of nutrients to grow and reproduce. We do not know how many survive the trip, but we know that a significant fraction does: They spread this way.

At their destination, the spores germinate to grow new vegetation; the bacteria multiply; the viruses stay dormant until they find a host; and the dust settles into a geological stratum, fertilizes a crop, or occasionally annoys car owners in the form of muddy rain. Whole ecosystems, including cities, become influenced through rainfall, snow, or fog by distant life forms, foreign silts, and pollution.

The surprise here is not so much that bacteria and viruses are so resilient but that they—and presumably, other life forms as well—appear to have hijacked a microfluidic transportation mechanism that predates the existence of life itself—one that Nature seemingly provided for the more planetary purposes that we see manifested in rain, clouds, and the rainbows that existed well before poets and scientists such as Newton paid attention to them. That's billions of years of evolution looking at you. By monitoring global weather patterns, we have learned to appreciate that Nature spreads the kernels of life inside tiny droplets, the microfluidic pearls of our planet.

Summary

- The rainbow is an optical phenomenon caused by *raindrops*—microfluidic drops suspended in the air—that are acting as miniature prisms, as correctly surmised by Newton in 1665.
- Raindrops form by *condensation* around *aerosols*—micrometer-sized airborne particles of mineral or biological origin.
- Aerosols and bioaerosols can have an enormous impact on Earth's climate and life.

2

The Drops of Life

The Role of Surface Tension in Everyday Life—and Biotechnology

Water, water, everywhere,
And all the boards did shrink;
Water, water, everywhere,
Nor any drop to drink.
> —Samuel Taylor Coleridge (1772–1834), in "The Rime of the Ancient Mariner"

Life on Earth would not exist without the wondrous propensity of water droplets to hold their shape, as you see them rolling on a leaf or your windshield. By this measure, a sailor named Steve Callahan is doubly indebted to these life-sustaining microfluidic systems. In 1982, when he was thirty years old, Callahan was able to survive seventy-six days on a life raft adrift in the Atlantic Ocean thanks to the solar still,[1] an ingenious droplet-generating desalination device invented approximately forty years earlier by Massachusetts Institute of Technology (MIT) professor Mária Telkes[2] (Figure 2.1, top).

Telkes was born in Hungary and was spellbound by solar energy from an early age. She studied physical chemistry in Budapest and emigrated to the United States. There she became an MIT professor and one of the founders of solar thermal storage systems, earning her the nickname "the Sun Queen." For her many practical inventions, such as a solar oven, the solar still, and the first modern residence heated by solar energy, Telkes was inducted into the National Inventors Hall of Fame.[2] During World War II, Telkes was called to serve in the U.S. Government's Office of Scientific Research and Development. Her solar still, commissioned by the U.S. Navy during that period, saved countless lives of

---

[1] Williams, D. For 76 days this man was adrift in the Atlantic—survived by using a solar still. *Outdoor Revival.* https://www.outdoorrevival.com/news/for-76-days-this-man-was-adrift-in-the-atlantic-survived-by-using-a-solar-still.html?firefox=1 (2016).
[2] Rafferty, J. P. Mária Telkes: American physical chemist and biophysicist. *Encyclopedia Britannica.* https://www.britannica.com/biography/Maria-Telkes (2021).

Dec. 10, 1968   M. TELKES   3,415,719
COLLAPSIBLE SOLAR STILL WITH WATER VAPOR PERMEABLE MEMBRANE
Filed May 11, 1966

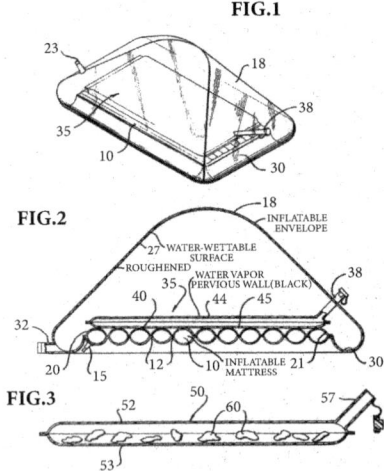

Figure 2.1 (Top) Mária Telkes in 1956. (Bottom) Illustrations from the solar still patent by Mária Telkes.

Image sources: (Left) Library of Congress. Wikipedia. https://en.wikipedia.org/wiki/ M%C3%A1ria_Telkes#/media/File:Maria_Telkes_NYWTS.jpg. (Right) Google Patents, US3415719A. https://patents.google.com/patent/US3415719A/en.

downed airmen and torpedoed sailors[3] during the war, and it has since become a critical piece of equipment in every mariner's survival kit.

In the fall of 1981, Steve Callahan sailed with a friend from Bermuda to England on a 21-foot sailboat, the *Napoleon Solo*, which Callahan had designed and built himself. Their trip went as planned, and they arrived off the coast of England in about two weeks. Callahan planned an easy solo return to the United States via Antigua. He was highly experienced and well prepared, and the wind would be at his back. He set off on his trans-Atlantic trip from the Grand Canary Islands on January 29, 1982, without the slightest premonition that within a week he would be fighting for his life for the next two and a half months on a 6-foot-diameter inflatable life raft with the few things he could salvage from his sinking boat—what he later described as his "view of heaven from a seat in hell."[4]

In the middle of the night of February 4, while sailing through a gale in the mid-Atlantic, *Napoleon Solo* was violently struck by what Callahan believed to be a whale. Water began rushing in through a 4-foot gash in the hull, and Callahan immediately realized that his boat would sink. Overcoming his panic, he instinctively put his training and experience to work. He had read books by other adrift survivors, such as Robertson,[5] about events that occurred before his trip. He quickly launched the self-inflating life raft into the churning sea and tethered it to the side of the sailboat. To salvage his most vital emergency gear, he dove several times into the flooding hold by the full moon's light. By the time he cut the life raft free and saw his boat drift away, he had managed to salvage some cans of food, a speargun, a survival kit, utensils, navigation instruments, a flare gun, and three Telkes solar stills—without which, despite having *water, water everywhere*, he would not have had a *drop to drink*.

Along with his strong determination and constant vigilance, the Telkes devices provided just enough water for Callahan to survive. These solar stills are based on the evaporation of seawater and the drop-by-drop condensation of distilled water.[6] It took some time and tinkering for Callahan to get the stills up and running. They didn't always function optimally under the unstable sea conditions he encountered over the four months he was adrift. However, for the most part, the stills delivered about a half liter—approximately 12,000 drops—of potable water a day.

Once inflated, a Telkes solar still looked like a large, transparent, rectangular balloon closely tethered to the side of the life raft (Figure 2.1, bottom). A porous black pad on the bottom of the still allowed seawater to seep into the balloon as it floated along the ocean surface. The Sun's rays heated the black surface and elevated the temperature inside the balloon, accelerating the rate

[3] National Inventors Hall of Fame. Mária Telkes: Solar thermal storage systems. https://www.invent.org/inductees/maria-telkes (n.d.).

[4] Kraken Yachts. A view of heaven from a seat in hell. https://krakenyachts.com/a-view-of-heaven-from-a-seat-in-hell (2021).

[5] Robertson, D. *Sea Survival: A Manual*. Praeger (1975).

[6] Callahan, S. *Adrift: Seventy-Six Days Lost at Sea*. Houghton Mifflin (1986).

Figure 2.2 The shape of a drop is stabilized by molecular forces.
Image source: Pxhere.com. https://pxhere.com/en/photo/1337421.

at which water naturally evaporates. The evaporated water left the saltwater behind. Meanwhile, the balloon's surface—transparent and in contact with the outside cold air—remained cool, causing the water vapor inside the balloon to condense onto the interior walls. This condensation formed drops of salt-free water that naturally beaded up, in a process that mimicked the formation of dewdrops on the surface of leaves in a chilly morning (Figure 2.2).

The beading up of water drops that saved Callahan's life is a prime example of how life critically depends on the peculiar microscopic properties of water. Water molecules strongly attract each other—more than in any other liquid except mercury. The water molecules on the surface of a pond are not like the water molecules deep in the pond. The surface molecules hold on tightly to their surface neighbors because, being exposed to air, they lack water molecules above them. Inside the liquid, on the other hand, cohesive forces attract the rest of the molecules to each other. As a result, a net inward force causes water to behave as if a stretched elastic membrane were covering the water's surface. This force, called *surface tension*, allows water striders to skim over the surface of lakes (Figure 2.3). It is also responsible for the spherical shape of the water droplets suspended in the air, acting as tiny lenses to project rainbows in the sky (see Chapter 1). Without surface tension, we would not have raindrops or dew, and Telkes knew that very well.

Surface tension was a key feature of Telkes' design. She found that a water-repellent ("hydrophobic") plastic liner she devised helped the water roll down the walls more efficiently. Once the droplets achieved a specific size, gravity pulled them down, one by one, into a gutter-shaped reservoir at the device's base. Callahan could drink the life-saving desalinated water from a small spout when the reservoir was full.

Figure 2.3 Water striders can stay and move on the surface of water due to a combination of their small weight, their water-repellent legs, and the high surface tension of water.

Reproduced from Corey (2006). Gerrini nymphs in Higashitakane Shinrin-koen (forest park), Miyamae, Kawasaki, Kanagawa, Japan. Wikipedia. https://en.m.wikipedia.org/wiki/ File:Amenbo_06f5520sx.jpg#file. Under a Creative Commons Attribution-Share Alike 2.1 Japan (CC BY-SA 2.1 JP).

The flare gun was useless—he spotted seven large boats in the distance, but the flare did not attract them. On the other hand, the speargun came in handy when he ran out of canned food. He could fish around his lifeboat from the school of large dorados that formed part of a busy ecosystem (including barnacles and sharks). His navigation instruments reassured him that the currents were taking him toward his intended destination. Finally, attracted by the birds around his raft, fishermen rescued him about 3,000 kilometers from his shipwreck, just south of Antigua. He had lost one-third of his weight and was tired of eating dorados. Still, he had not died of thirst—thanks to a distilling device that had vaporized the seawater around him into drinkable water drop by drop by drop.

* * *

Scientists have continued to build on the work of Telkes, and other pioneers in early microfluidic applications have tried to put the surface tension of droplets to various uses. One area in which this has been particularly fruitful is in scientific labs. Biology experiments are expensive and tedious and need to be repeated over and over again. Conjure up in your mind for a second a picture of the "mad scientist" in their lab, surrounded by hundreds of test tubes and beakers filled with bubbling liquids—now imagine each reaction going on in this picture reduced to a different, tiny droplet. Miniaturizing the whole experiment would reduce the cost and would allow for replicating the experiment 100-fold, perhaps 1,000-fold, in the same amount of time that you take to drink a cup of coffee or scroll through your favorite social media feed.

But *manipulating* the contents of each droplet can be tricky, as David Beebe, a professor of bioengineering at the University of Wisconsin, can attest. Beebe is a sporty bioengineer from West Salem, a small Wisconsin farming town near the Mississippi River, who trains for athletic events—serious ones, like 10-kilometer road races and triathlons—in his free time. Because I have known Dave for the better part of my microfluidics career, I have witnessed the many tricks this Houdini of surface tension has pulled from his hat throughout the years. At his University of Wisconsin lab, he has invented several microfluidic devices that induce droplets and fluids to flow in a seemingly autonomous way based on his deep understanding of the physics of fluids at small scales. He owes his passion for engineering to growing up in a "very mechanical environment." His father was an agricultural engineer and owned a farm equipment dealership during Beebe's formative years. "So I worked on tractor engines and helped my grandfather fix stuff on the farm," he recalled. Halfway through grad school, he read "Life at Low Reynolds Number."[7] This paper is the transcript of an unassuming yet visionary and influential talk given in 1977 by the American physicist and Nobel Laureate Edward M. Purcell. In this talk, Purcell explained the forces and turbulence-free effects that flow has at small scales. "I was hooked on the physics of the microscale!" Beebe said.

---

[7] Purcell, E. M. Life at low Reynolds number. *Am. J. Phys.* 45: 3–11 (1977).

Beebe's devices have elegantly simple designs that borrow sophisticated technology from the microelectronics industry. One of my all-time favorite Beebe designs couldn't be simpler: It consists of a tiny straight channel—as wide as a hair and a few centimeters in length—connecting two reservoirs. To make the device, his group's students first prepare a mold from which they replicate the final device. Because a speck of dust can do the same damage to the channel's mold as a boulder falling to an interstate road, they gown up in outer-space-like suits and enter a dust-free lab called a *cleanroom* to fabricate the mold. In the cleanroom, they use a light-based technique similar to photography called *photolithography,* developed initially to make small circuits in microelectronic chips (Figure 2.4).[8] The astronaut–student typically covers a substrate with a

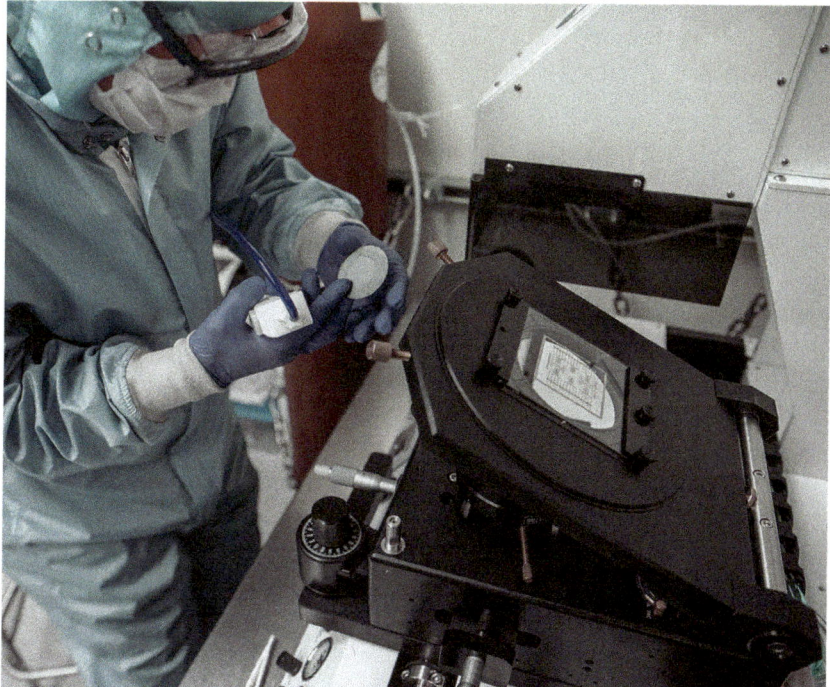

Figure 2.4 This image, taken inside a cleanroom, shows one step in the process of photolithography. An operator, fully gowned to protect the devices from dust, places a silicon wafer with his left hand inside an aligner (black apparatus), where it will be exposed to UV light through a photomask. The photomask is already bolted on the aligner. The silicon wafer has been coated with a thin layer of photosensitive resin called *photoresist.*
Image courtesy of the University of Washington.

[8] Xia, Y. N., & Whitesides, G. M. Soft lithography. *Angew. Chem. Int. Ed. Engl.* 37: 550–575 (1998).

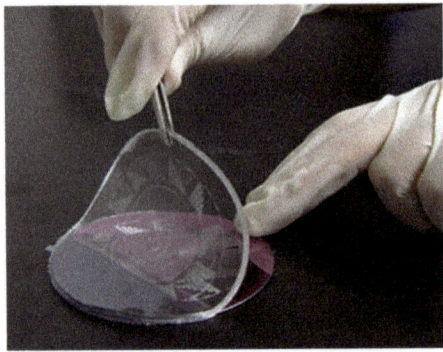

Figure 2.5 A student peels the transparent rubber replica away from the photolithographically patterned master mold (on a silicon wafer). This simple procedure—also known as *soft lithography*[8]—is commonly used to fabricate microfluidic channels.

Image courtesy of Gang Li, School of Optoelectronic Engineering, Chongqing University, Chongqing, China.

photosensitive polymer and shines ultraviolet (UV) light through a *photomask*, which only lets through the desired pattern. The areas exposed to UV become insoluble to a chemical, and the unexposed regions dissolve away, creating the mold. Finally, the student casts the replica in a transparent rubber material, separates it from the mold (Figure 2.5), and bonds it to a flat substrate to finish the device, which can safely exit the cleanroom with the student. This old technique is still used today to fabricate most microfluidic devices.

In 2002, Beebe asked his student Glenn Walker to put cells in the channel of one such straight-channel device and run a routine experiment that became revelatory. In the lab, cells need to be fed, and their environment needs to be cleaned up periodically. Inside our body, these routine maintenance functions are performed by the blood's circulatory system, which is powered by the heart. In their experiments, scientists have to find ways to move fluids in the absence of a heart. Walker had devised a "blood-like" method to exchange the cellular fluids by inserting a pipette tip into one of the reservoirs and squirting fluid into the channel. This protocol produced a large droplet at the other reservoir, which he then aspirated with a pipette. The main problem with moving fluids with pipettes is that it is very tedious and requires a lot of manual dexterity (Figure 2.6).

One day, Walker accidentally made a pipetting error and placed a droplet *on* the reservoir rim instead of *in* it. To his surprise, the droplet immediately disappeared. Walker realized that something interesting had just occurred and,

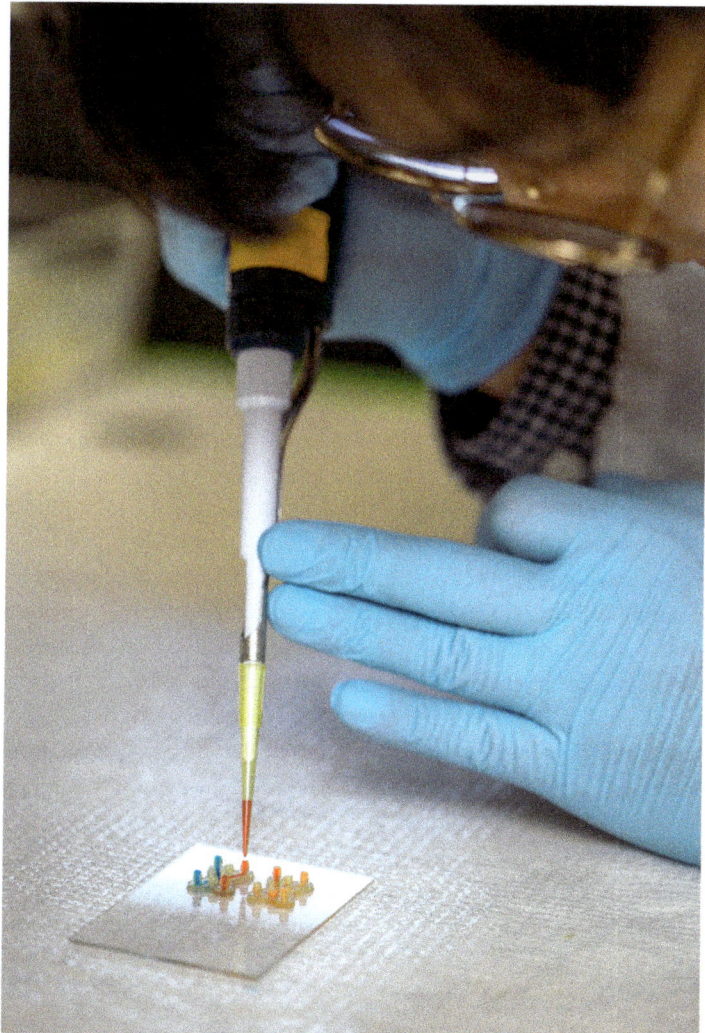

Figure 2.6 Pipetting is a very common procedure in biomedical laboratories. It is very tedious and prone to human error. The researcher in the image is using a pipette to introduce fluids in a microfluidic device.
Image courtesy of the University of Washington.

together with Beebe, realized the observation was precisely what one would expect from the physics of a tiny drop.

Many engineers had used similar devices before them. Still, they had only placed a drop on one end, resulting—as expected—in the drop wicking into the channel by capillarity. What Walker did differently this time was add one drop on *both ends* and of different sizes. They had expected the large drop to "apply weight" and flow toward the small one until they equilibrate, like a cask of wine emptying when its spigot is open. Counterintuitively, the large drop grew in size, drawing fluid from the small drop—as if the wine barrel had refilled itself against gravity upon opening the spigot!

Walker and Beebe showed that this fluid movement is governed by surface tension, independently of gravity. In other words, the height and weight of the drops play a minimal role. The surface tension is inversely proportional to the radius and thus is stronger in the smaller drop.[9] The smaller a drop is, the stronger its internal pressure and the harder it can pump; even though it can only pump for a short time, it is a handy, "tubeless" pump.[10] None of this happens at the scale of a cask of wine, where the surface tension is negligible compared to the enormous weight of the wine inside.

The absence of tubing is essential here. Using these kinds of stunts, Beebe's group has run nearly 100 miniature experiments in parallel by running arrays of these channels filled with cells (Figure 2.7).[11] The platform allows for feeding the cells simply by dispensing drops with standard equipment on the surface of the device—no tubes needed! These magic tricks are invaluable to biologists and clinical scientists who need to perform myriad experiments with small quantities of expensive reagents and do not want to be bothered with complex instrumentation. The device could be used to design new easy-to-use personalized treatments, for example, by testing how certain drugs affect a given patient's cells once these are introduced into the channels. Beebe's need for simplifying a biotechnology assay is not unlike the need for collecting portable water, a most scarce reagent in the middle of the ocean, that motivated Telkes' solar still design which saved Callahan's life. Drop by drop, surface tension can help save human lives.

---

[9] Walker, G. M., & Beebe, D. J. A passive pumping method for microfluidic devices. *Lab Chip* 2: 131–134 (2002).

[10] Berthier, E., & Beebe, D. J. Flow rate analysis of a surface tension driven passive micropump. *Lab Chip* 7: 1475–1478 (2007).

[11] Meyvantsson, I., Warrick, J. W., Hayes, S., Skoien, A., & Beebe, D. J. Automated cell culture in high density tubeless microfluidic device arrays. *Lab Chip* 8: 717–724 (2008).

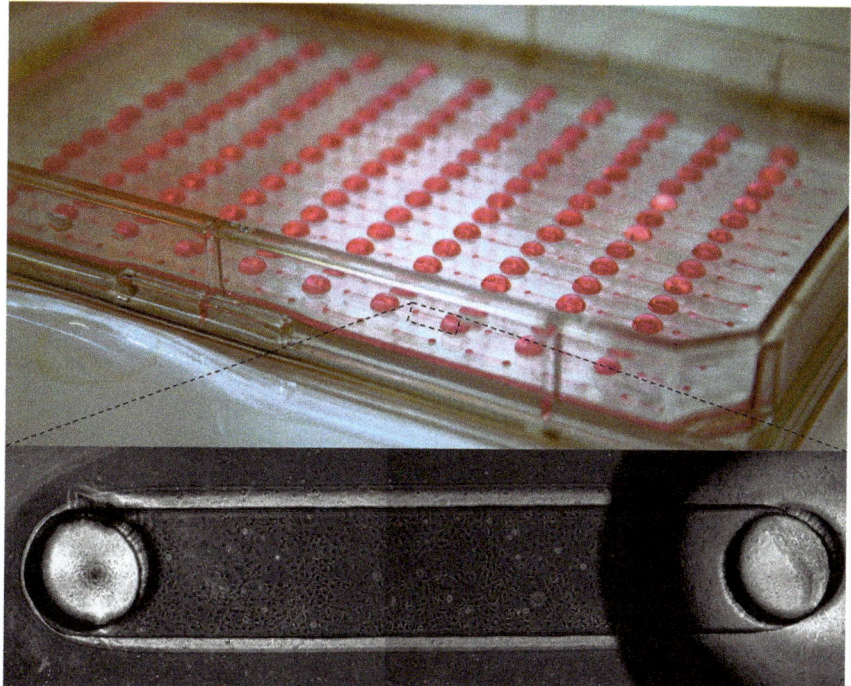

Figure 2.7 Arrays of microchannels filled with cells. Each channel between drops is 4.5 centimeters long, and the drops at the two ends serve to pump fluids using surface tension.

Images courtesy of Ivar Meyvantsson and David Beebe.

Summary

- A solar still is a desalination device invented by MIT professor Mária Telkes that relies on the condensation of water droplets and their surface tension to produce fresh water in the middle of the ocean.
- Microfluidic devices that manipulate droplets can be useful for generating many biological measurements in parallel.
- The first step in the fabrication of a microfluidic device is often photolithography, a process similar to photography that takes place in a dust-free cleanroom.
- The result of photolithography is typically a patterned flat substrate that is used as a mold to fabricate multiple microfluidic devices.

3

Healing Mists

The Breakup of Jets into Sprays and Their Use in the Treatment of Tuberculosis and Asthma

Like many people debilitated by asthma, particularly professional athletes, world record holder Jacqueline Joyner-Kersee can attest that microfluidic nebulizers changed her life (Figure 3.1, top). Asthma is a condition in which an allergy or infection causes the muscles surrounding the bronchial tubes—the passages or airways that allow air to enter and leave the lungs—to constrict, making breathing extremely difficult and possibly even causing death.

An asthma attack—which can last from minutes to days—can be a frightening experience. It can feel as if a person is sitting on your chest or as if you are drowning in air. You-struggle-to-take-a-full-breath. Yourbreathingquickens.

Joyner-Kersee, one of the all-time greatest track-and-field athletes, was born in 1962 and excelled in both track and field and women's basketball while attending college at the University of California, Los Angeles (UCLA) from 1980 to 1985. In 1983 and 1985, she won the Broderick Award, given to the nation's best female collegiate track and field competitors. In April 2001, Joyner-Kersee was elected the "Top Woman Collegiate Athlete of the Past 25 Years."

During this period, as a top student-athlete at UCLA, Jackie was diagnosed with asthma. Because she was afraid her coaches would make her stop running, she did not tell them about her condition at first. "I was always told as a young girl that if you had asthma, there was no way you could run, jump, or do the things I was doing athletically," she said. "So, . . . it took me a while to accept that I was asthmatic."[1] Alongside her and soccer superstar David Beckham, asthma affects 300 million people worldwide, including 10% of Americans—numbers that increase by 50% every decade.[2]

Microfluidics came to Joyner-Kersee's rescue by way of a muscle relaxant delivered directly to her trachea in the form of a mist with an *inhaler*. This hand-actuated device uses a miniature nozzle to spray medicines as tiny droplets into the mouth (Figure 3.1, bottom). The mist sprayed by the inhaler immediately

[1] Jackie Joyner-Kersee: Living with Asthma. *NIH Medline Plus—The Magazine(Fall 2011)*. https://magazine.medlineplus.gov/pdf/MLP_Fall_11.pdf (2011).
[2] Braman, S. The global burden of asthma. *Chest* 130: 4S–12S (2006).

Figure 3.1 (Top) Jackie Joyner-Kersee in the 1988 U.S. Olympic trials. That day, she set a long jump record. (Bottom) An inhaler for asthma patients similar to the one Joyner-Kersee used to be able to compete.

Image sources: (Top) Wikipedia. https:// en.wikipedia.org/wiki/Jackie_Joyner-Kersee#/media/File:Jackie_Joyner-Kersee_1988b.jpg. Public domain. (Bottom) Pxhere.com. https://pxhere. com/en/photo/660816.

reverses the asthma attack.[a] Joyner-Kersee said, "It took me a while to even start taking my medication properly, to do the things that the doctor was asking me to do. I just didn't want to believe that I was an asthmatic. But once I stopped living in denial, I got my asthma under control, and I realized that it is a disease that can

---

[a] The medication in an asthma inhaler can be classified as either a "reliever" or a "preventer." An example of a reliever drug is albuterol (commonly marketed as Ventolin), which causes the airways to widen by relaxing the muscles in the lungs, thus reversing asthma attacks in a short time. An example of a preventer drug is budesonide (marketed under the trade name Pulmicort). Preventer drugs are typically taken daily. They act to reduce inflammation in the long term and do not provide relief if you are already experiencing an asthma attack.

be controlled."[3] She won three Olympic and four World Outdoor Championship gold medals, and her heptathlon world record of 7,291 points obtained at the 1988 Summer Olympics in Seoul is one of the longest standing sports records. In the same Olympic Games, she was the first American woman to earn a gold medal in long jump and the first American woman to earn a gold medal in heptathlon. *Sports Illustrated* chose Joyner-Kersee as "The Greatest Female Athlete of the 20th Century."

Inhalers were not new when Joyner-Kersee started using them. The first spray-generating devices date back to the mid-1800s. They were conceived in response to the tuberculosis epidemic of the nineteenth century—coinciding with the Industrial Revolution and the mechanical engineering required to make those devices—and in the ancient belief that water itself had curative properties.[b] Tuberculosis was then and still is a highly contagious disease caused by the rod-shaped bacterium *Mycobacterium tuberculosis*. The disease spreads via the respiratory route when an infected person talks, coughs, sneezes, or exhales minute droplets containing the bacteria, which another person inhales. Infection triggers a slow decline characterized by loss of body weight, depression, pale skin, and sunken eyes. Until antibiotics were developed after World War II, tuberculosis had devastating effects across societies enmeshed in the Industrial Revolution of the eighteenth and nineteenth centuries. The disease swiftly spread through crowded cities, where workers and families who had moved from the countryside for factory jobs often lived in close, unsanitary quarters. Known at the time as "white plague" or "consumption," tuberculosis had a heavy, lasting impact on society; it was a recurrent theme in the arts spanning over three centuries.[c] In Europe alone, the disease caused one-quarter of all deaths.[d]

In the absence of effective remedies against this plague, as doctors desperately pursued any conceivable method that could lead to treating ailments of the lungs and airways, they looked to water. Water had been considered as a treatment since Antiquity, as water could not be harmful anyways. Both Aretaeus and Galen, prominent Greek medical researchers who lived in the second century AD, had concluded that the salty mist produced by the breaking of waves had

---

[b] The idea that respiratory ailments could be cured by inhalation dates back to the Egyptians, but it was limited to smoke or vapors.[3]

[c] Tuberculosis appears as a theme in paintings by both Rembrandt (1642) and Monet (1879); Giacomo Puccini's opera *La bohème* (ca. 1893); Thomas Mann's novel *The Magic Mountain* (1924); and in the poignant song by Van Morrison, *T. B. Sheets* (1967)—to name a few examples.

[d] Prominent people who died of tuberculosis include English poet John Keats (d. 1821), U.S. President Andrew Jackson (d. 1845), Polish composer Frédéric Chopin (d. 1849), Russian playwright Anton Chekhov (d. 1904), German writer Franz Kafka (d. 1924), English author George Orwell (d. 1950), and First Lady of the United States and human rights activist Eleanor Roosevelt (d. 1962).

[3] Stein, S. W., & Thiel, C. G. The history of therapeutic aerosols: A chronological review. *J. Aerosol Med. Pulm. Drug Deliv.* 30: 20–41 (2017).

a favorable effect on the lungs.[4] The marble walls of Roman public baths often had the letters "SPA" on them—short for *Salude Per Aqua* or "health through water." in Latin. In the early 1840s, doctors started reviving the traditionally held notion that seaside sprays—airborne droplets or aerosols from crashing waves—could benefit the treatment of respiratory diseases. "Hydrotherapy" sessions gradually became part of the cure in sanatoria, dedicated high-mountain facilities where doctors sent patients to "take the waters."[e] In the nineteenth century, in Germany, Poland, and Austria, spa clinics for tuberculosis patients spawned around brine graduation towers—salt-producing plants consisting of wooden buildings that constantly trickle water, thus forming salty aerosols.[4]

In one of these spas at Euzet-Les-Bains in the south of France, Dr. Auphan wanted to introduce medicines into the lungs of tuberculosis patients who visited his facility. In 1847, inspired by the sprays created in waterfalls or at the seashore when water crashes violently against surfaces, as well as by the purported effects of salty mists, he invented a method for generating sprays by blasting jets of salt water against a surface at very high pressure, producing a cloud of droplets inside a so-called vaporatorium. He argued that the *atomization* of fluids into droplets was a more effective strategy for introducing mineral waters into the lungs than the vapors from aromatherapy of the time.[f] Many spas throughout Europe rapidly adopted Auphan's atomization approach.

Auphan's approach used plentiful amounts of fluids, so doctors could not implement it next to a patient's bed. In 1858, a little over a decade after Auphan's invention, in Pierrefonds, near Paris, Dr. Jean Sales-Girons developed a "pulvérisateur" or portable atomizer, a nozzle that sprayed droplets directly into the throat of patients (Figure 3.2).[5] Sales-Girons' "pneumatic atomizer" device was the miniature version of Auphan's method: A compact

---

[e] Sanatoria were moderately effective against tuberculosis, but it is unlikely that the various treatments involving water—typically from mineral-rich geothermal sources—had any curative effect on the patients, who were also treated to abundant rest, fresh air, sunshine, and good nutrition. These unsupported health claims of water were not unique to tuberculosis treatments. Indeed, the idea that hydrotherapy has curative effects persists in many countries and communities. Germany alone has more than 100 spa towns whose names start with *Bad* ("bath") or contain Baden ("to bathe") and a whole state named Baden-Württenberg. Until recently, German insurance would pay for "kur" ("cure")—six weeks of spa treatments—to recover from surgery.

[f] Hippocrates' pot-and-reed inhaler design had been used since the fourth century BC, and the British physician and astronomer John Mudge had invented an inhaler in 1776 based on a pewter tankard.[6] These devices were generally designed to concentrate vapors from substances such as opium and were used to treat asthma or cough, but they had failed to be effective against tuberculosis.

[4] Nerbrink, O. A history of the development of therapy by jet nebulisation. In *Optimization of Aerosol Drug Delivery* (ed. L. Gradon & J. Marijnissen), 1–22. Kluwer Academic, 2003. doi:10.1007/978-94-017-0267-6_1.

[5] Sales-Girons, J. *Thérapeutique Respiratoire—Instruction sur l'Instrument Pulvérisateur des Liquides Médicamenteux, ses Applications au Traitement des Maladies de Poitrine et la Manière de s'en Servir.* Charriere, 1861.

[6] Anderson, P. J. History of aerosol therapy: Liquid nebulization to MDIs to DPIs. *Respir. Care* 50: 1139–1149 (2005).

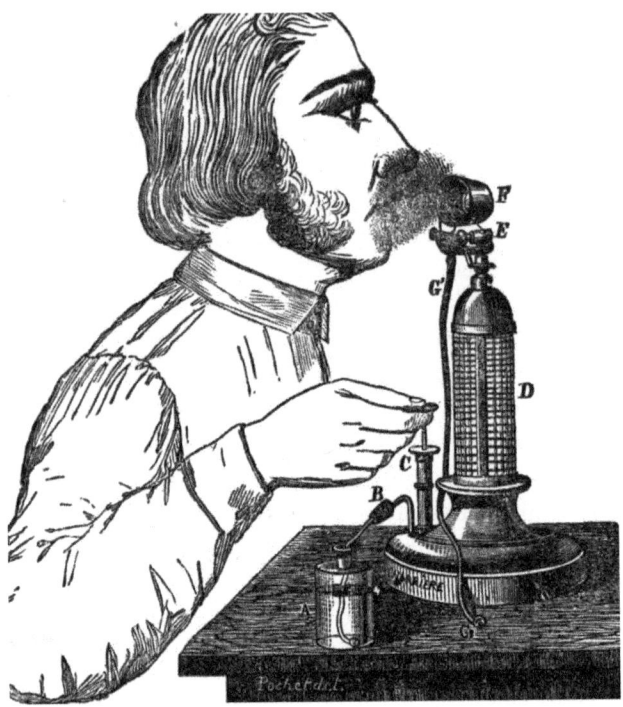

Figure 3.2 Sales-Girons' "pulvérisateur." The hand-operated pump (C) projects a "capillary jet" (E) through an orifice in a cone (F), inside which it produces a spray when the jet collides with a metal disk.

Image source: Sales-Girons' "pulvérisateur," from an illustration in his own 1861 book.[5]

hand-operated pump generated a large pressure[g] to shoot a tiny fluid "capillary jet" at high speed against a metal disk, producing a spray of droplets (see Figure 3.2). The patient was then asked to open their mouth so the doctor could direct the spray into their gorge.

In Sales-Girons' time, scientists were beginning to understand the physics of how a water stream evolves in mid-air. Two scientists went on to explain the underlying principles of Sales-Girons' device. The first was the Belgian physicist and mathematician Joseph Plateau, who observed in 1873 that liquid streams break into droplet trains that are too fast to be visible to the naked eye. Why are water streams in the air, like a water fountain (Figure 3.3), so unstable, but oil streams are much steadier? The British physicist John William Strutt, 3rd Baron Rayleigh (known as Lord Rayleigh[h]), provided a straightforward explanation a

[g] Sales-Girons' pump could generate a pressure of 3 atmospheres. By comparison, a soccer ball is inflated to just above 1 atmosphere of pressure.

[h] Lord Rayleigh (1842–1919) received the 1904 Nobel Prize in Physics (for the discovery of argon) and made extensive contributions in a wide variety of scientific fields, such as optics (his

few years after Plateau's observation: As soon as the water exits the pipe and loses the support or wetting contact from the pipe wall, the surface tension of water is so high that it starts slicing the stream like a loaf of bread.[7] You have seen it many times in your dripping faucet or showerhead. As the water tries to minimize its surface energy, it forms the most stable of shapes: spheres. In contrast, you will never see the stream of olive oil coming out of the oil cruet or pouring from a spoon breaking into droplets because oil has a much lower surface tension. Now known as the Plateau–Rayleigh instability, or *Rayleigh instability* for short, this principle governs the formation of droplets in all pressure-powered (or "pneumatic") jets. Many atomizers, water sprinklers, and fountains use the Rayleigh instability.

The efforts of Plateau and Rayleigh were not isolated to the effervescent field of fluid mechanics. About a century and a half before, a prominent eighteenth-century Swiss mathematician named Daniel Bernoulli had studied what happened when someone connected a hose to the bottom of a tank through a faucet. He applied the principle of conservation of energy, which states that the total energy of a closed system is always constant and conserved, although it can transform from one form of energy to another.[i] At the faucet, inside the tank, the pressure was higher than atmospheric pressure due to the weight of the water in the tank. When he opened the spigot, the fluid started at rest in the tank. It gradually accelerated due to the pressure difference until it exited the hose at high speed: The liquid had moved from a state of zero velocity and high pressure to a state of high speed and atmospheric pressure. From this observation, Bernoulli inferred in 1738 that the highest speed always occurs where the pressure is lowest, and the lowest velocity occurs where the pressure is highest. He did not formulate it mathematically, but he had stated one of the foundational principles of fluid mechanics—that pressure inside a pipe decreases when the flow speed increases, and vice versa. Now named *Bernoulli's principle*, it is valid at all known scales, from rivers to microfluidic channels.[j]

explanation of why the sky is blue is known as Rayleigh scattering), theory of sound, and fluid dynamics (aerodynamic lift), among others.

[i] A classical example of energy conservation is observed in the drop of a rigid ball from a certain height. Before you drop the ball, it has a certain *potential* energy—the energy that an object possesses by virtue of its height above the ground. Once the ball is dropped, it immediately begins losing potential energy and gaining speed or *kinetic* energy—the energy that an object possesses by virtue of its motion. Potential and kinetic energy have an inverse relationship because the total energy remains constant: As one decreases, the other one increases in exact amount. At the moment before the ball hits the ground, its kinetic energy is equal to the potential energy it possessed before it was dropped. Correspondingly, its potential energy is zero—the same amount of kinetic energy it possessed before it was dropped.

[j] The reason Bernoulli's principle is valid at all scales is precisely because it is simply the particular expression for fluids of the universal principle of conservation of energy.[8]

[7] Rayleigh, F. R. S. On the instability of jets. *Proc. London Math. Soc.* 10: 4–13 (1878).

[8] Vogel, S. *Life in Moving Fluids: The Physical Biology of Flow.* Princeton University Press, 1994.

Figure 3.3 High-speed photography of a water fountain reveals the breakup of the streams into droplets. This breakup is imperceptible to the naked eye and is the principle behind the Sales-Girons nebulizer and many other devices.

Image source: Pxhere.com. https://pxhere.com/en/photo/998378.

Luckily for future generations seeking practical uses of Bernoulli's principle, Daniel's father was one of the foremost mathematicians of the time. One of his trainees, Leonhard Euler, who would become one of the greatest mathematicians in history, was a close friend of Daniel's and put Bernoulli's principle into mathematical form in 1752. Formulas are essential because they allow engineers to conceive exact designs. For example, when hydraulic (or microfluidic) engineers are asked to design a system of pipes that distribute a stream equally among a given number of tubes, they apply Euler's version of Bernoulli's principle. The principle also predicts why wine flows faster when we decant the bottle more, when the cask is fuller, or when the spigot is opened wider.

Giovanni Battista Venturi, an Italian physicist, was interested in examining this phenomenon by asking a different question: What happens to fluids when forced through a constriction? The question still baffles present-day students because the answer is not intuitive. Do liquids slow down and "feel" more pressure—as one would imagine from a crowd going through a door—or the other way around—as happens to a river when it reaches a narrowing and accelerates? To investigate, in 1836, Venturi built a device consisting of a fluid pipe that incorporated a constriction and two orthogonal fluid pipes for measuring pressure (Figure 3.4). Venturi observed that the device—somewhat counterintuitively but confirming Bernoulli's principle[k]—generated a point of *high speed* and *low pressure* at the constriction.

Similar spray devices were invented shortly after Sales-Girons' 1858 portable atomizer. In 1859, an inventor named L. Mathieu exhibited his *Néphogène* spray device to the Academy of Medicine in Paris.[9] And in 1862, the German physician Bergson developed the *Hydrokonium*. Remarkably, the Sales-Girons spray utilized the Rayleigh instability while the Néphogène and the Hydrokonium atomizers used the Venturi trick, but in all three cases, the user only saw and felt the same spray. Mathieu designed the fluid to drip through a tube and fall perpendicularly into a pressurized air stream. The collision between the falling fluid and the air stream generated a spray of droplets. Bergson's Hydrokonium also used air blown through a side pipe that had a constriction. Due to the Venturi effect, when he blew air through the tube, the reduction in pressure at the

---

[k] This observation runs counter to intuition because to generate higher flow speeds (i.e., to increase it *in time*) we do require higher head pressures to drive the flow. However, the pressure and the flow rate change within the pipe. What Bernoulli's principle says is that comparing *any two points in a pipe* with flow running *at any given head pressure*, the point with higher speed of flow (an increase *in space*) is at lower pressure.

[9] Solis Cohen, J. *In the Treatment of Disease: Its Therapeutics and Practice. A Treatise on the Inhalation of Gases, Vapors, Fumes, Compressed and Rarefied Air, Nebulized Fluids, and Powders.* Lindsay & Blakiston, 1876.

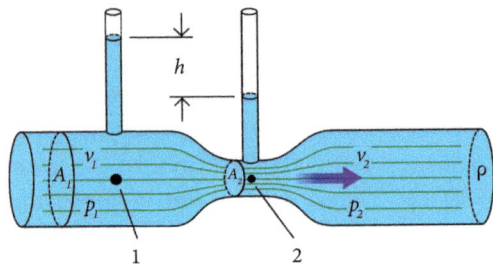

Figure 3.4 Schematic diagram of a Venturi device showing the flow lines (green). As the pipe's cross-sectional area narrows from A_1 down to A_2, the flow speed *increases* from v_1 to v_2 and the pressure *decreases* from p_1 to p_2. Two vertical pipes are used to measure the difference in pressure at those two points, shown here by the difference in the fluid height h.

Image source: User HappyApple on Wikipedia. https://en.wikipedia.org/wiki/Venturi_effect#/ media/File:Venturi5.svg. Public domain.

constriction caused fluid to rush toward a hole connecting with the air duct. As a result, fluid collided with the air current, generating a spray similar to Mathieu's.

Although we now take all of these spraying devices for granted, they caused a huge sensation at the time. It was the first time people saw a liquid turned into a "cloud" (*nebula* in Latin) through a machine. Hence, the device was termed a *nebulizer*, a synonym for atomizer. Dr. Solis-Cohen, a pioneering pulmonologist from Philadelphia, learned about the Sales-Girons device and started using it on his patients in the United States. One day, he was operating the Sales-Girons nebulizer for a patient when his friend exclaimed, "Well, Doctor, I'll be hanged! That's the first time I ever saw anyone get up steam out of cold water!"[9] Later studies by doctors concluded that both the water and the contents of the droplets had penetrated deep into the lungs.[4,9]

None of these atomizer designs ever cured tuberculosis, but they found many uses later on. On March 24, 1882, German physician Robert Koch appeared before the public with microscope slides at the University of Berlin's Institute of Hygiene to announce that the cause of tuberculosis was the *Mycobacterium tuberculosis* bacterium—for which he was awarded the Nobel Prize in 1905. The discovery of the microbial origin of tuberculosis led to the development of effective vaccines in the 1920s and antibiotic treatments in the 1940s. The Sales-Girons' sprays—or the high-altitude airs—had only relieved the symptoms caused by the tiny microbe. They never cured the patients.

It was not all in vain. The second half of the nineteenth century saw unprecedented inventions in mechanical and electrical devices feeding back on the economy and vice versa. Wondrous devices such as the telegraph, the telephone, the phonograph, the automobile, the cinema, and the radio were invented within the sixty years spanning from the 1830s until the 1890s.

Figure 3.5 (Top) Airborne droplets generated by a garden hose gun. (Bottom) Irrigation systems and sprinklers break streams of water into many droplets to improve the efficiency of water distribution.

Image sources: (Top) Pxhere.com. https://pxhere.com/en/photo/1294831. (Bottom) Pxhere.com. https://pxhere.com/en/photo/853379.

The availability of capital and the rapid development of better communications spurred information exchanges on new designs, which spawned many financial gains for the inventors—that is, more wealth. The Industrial Revolution generated a giant innovation snowball.

The sprays formed part of that snowball. In time, both spray designs (the Venturi-based spray and the pneumatic jet spray) would impact the treatment of many medical conditions—pulmonary infections, asthma, anesthesia, and cough—even though they had not been proven helpful for the tuberculosis patients they were initially intended for. Microfluidic sprays were all the rage at the end of the nineteenth century. They would generate many applications, from agriculture and sanitation to painting (see Chapter 6), cosmetics, and engines (see Chapter 4). Reviews of the time describe five atomizer designs in 1867[10] and twenty different ones in 1880, many commercially available.[11] In 1888, Dr. Allen

---

[10] Scudder, J. M. *On the Use of Medicated Inhalations in the Treatment of Diseases of the Respiratory Organs.* Moore, Wilstach & Baldwin, 1867.

[11] Beatson, G. On spray producers. In *The Glasgow Medical Journal* (ed. Coats, J.), vol. 14. Glasgow and West of Scotland Medical Association, 1880.

DeVilbiss, an otolaryngologist from Toledo, Ohio, started a successful company of atomizers that is still active today. The breakup of water jets into droplets is the basis of water hose guns and sprinklers (Figure 3.5), which have been used widely to water gardens, lawns, sports fields, and crops more efficiently for almost 100 years. Using sprinklers, farmers have been able to irrigate entire fields of vegetables in otherwise arid areas with mists of droplets and to spread pesticides over the crops. Public officials have sprayed disinfectants in municipal spaces, substantially impacting public health worldwide in the twentieth century.

The hand-operated pump of the Sales-Girons atomizer was ultimately superseded by more compact or portable forms of pressure generation, leading to less expensive, finger-actuated pneumatic sprays. Americans Lyle Goodhue and William Sullivan invented an insecticide spray can (called the "bug bomb") in 1947, the first *aerosol can*. This invention preceded—and possibly inspired—the first spray paint cans. The bug bomb allowed American soldiers deployed in the Pacific during World War II to defend themselves against malaria-carrying mosquitoes by spraying insect repellent inside tents.

In every aerosol can, a finger press opens the valve from a pressurized air chamber, generating a burst of pressure in the inner chamber of the container. This pressure burst ejects fluid in the shape of a jet through the nozzle at high speed; at high pressures, the jet's profile is closer to an onion or a tulip right outside the nozzle.[12] The breakup of the jet due to the Rayleigh instability generates the familiar cone of droplets of the spray can (Figure 3.6, top). When the finger presses the button, a valve opens inside the container, releasing pressurized gas, which causes the ejection of fluid at high speed (Figure 3.6, bottom). The liquid-containing chamber can also be pressurized via the displacement of a piston valve, as in a window cleaner spray bottle (Figure 3.7), perfume flasks (Figure 3.8), or a nasal spray (Figure 3.9). In all these devices, now fabricated in plastic, the fluid goes through a small microfluidic nozzle before contacting air, just like in the Sales-Girons device.

The efforts of Sales-Girons and other early spray developers—who envisioned the direct delivery of drugs in droplet form to the lungs—did bear their fruits. You might have used a commercial nebulizer in a modern hospital if you underwent surgery and the anesthesiologist covered your mouth and nose with a rubber mask connected to a spray-generating machine. The administration of inhalation anesthetics (e.g., isoflurane, desflurane, sevoflurane, or halothane, which are liquids at room temperature[1]), widely used to complement

---

[1] The gas nitrous oxide is also used as an inhalation anesthetic.
[12] Spray Analysis. Spray technology innovations in oil and gas. YouTube. https://www.youtube.com/watch?v=lWKd31jUABU (2021).

Figure 3.6 (Top) An aerosol can in action. (Bottom) The valve system has a stem that is part of the top actuator.

Image sources: (Top) User Andrew Magill on Wikipedia. https://commons.wikimedia.org/wiki/ File:Spray_can.jpg (CC BY 2.0). (Bottom) User Knulclunk on Wikipedia. https://en.wikipedia.org/ wiki/Aerosol_spray_dispenser#/media/File:Aerosol_tops_6.svg (CC BY-SA 3.0).

intravenous anesthetics in the operating room, often use nebulizers for their direct delivery to the lungs in the form of microfluidic droplets.[13]

As a star athlete in the 1980s, Jackie Joyner-Kersee likely used the portable version of a modern nebulizer called a *metered-dose inhaler*, introduced in the 1950s. In April 1955, George Maison, then president of Riker Laboratories,

---

[13] Sakai, E. M., Connolly, L. A., & Klauck, J. A. Inhalation anesthesiology and volatile liquid anesthetics: Focus on isoflurane, desflurane, and sevoflurane. *Pharmacother. J. Hum. Pharmacol. Drug Ther.* 25: 1773–1788 (2005).

Figure 3.7 A high-speed photograph of the droplets created by a spray bottle. The hand-operated pump ejects a water jet through the nozzle. After ejection, the jet breaks up into many droplets due to the Rayleigh instability.
Image source: Pxhere.com. https://pxhere.com/en/photo/1105796.

Figure 3.8 A manually actuated spray device. Here the jet is so unstable that, immediately after ejection, it forms a cone of droplets (the spray).
Image source: Pxhere.com. https://pxhere.com/en/photo/604280.

was inspired by his 13-year-old daughter Susie, who had severe asthma.[14] At the time, Riker was a subsidiary of Rexall Drug Company,[m] which produced hair spray. Susie asked him why the medication could not be contained inside a nice spray bottle like those used for hair spray. Maison thought this was a great idea and asked his team at Riker[n] to develop the device envisioned by his daughter. They used pressurized canisters and valves designed

---

[m] Riker is now called 3M Drug Delivery Systems.[14]

[n] Chemists Irv Porush and Charlie Thiel developed the first pressurized metered-dose inhaler at Riker.[14]

[14] Melling, L. Inventing the MDI: A history in modern inhalation therapy. Contract Pharma. https://www.contractpharma.com/contents/view_online-exclusives/2017-04-19/inventing-the-mdi-a-history-in-modern-inhalation-therapy (2017).

Figure 3.9 Nasal spray.

Image source: User robin_24 on Wikipedia. https://en.wikipedia.org/wiki/Nasal_spray#/media/ File:Action_photo_of_nasal_spray_on_a_black_background.jpg (CC BY 2.0).

for perfume glass flasks to deliver a short burst of aerosolized medicine (Figure 3.10). Riker rushed the U.S. Food and Drug Administration approval and launched the product at the end of March 1956, not even a year after Susie had candidly suggested it. For the first time, patients could reliably self-administer a specific amount of medication to the lungs using a metered-dose inhaler.

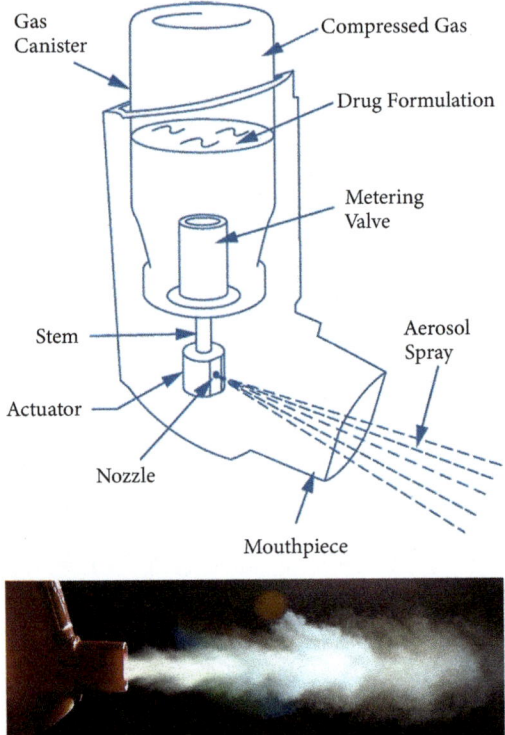

Figure 3.10 (Top) Schematic of a metered-dose inhaler. (Bottom) When the patient pushes down the gas canister, the metering valve opens and the pressure in the gas canister causes the drug to release at high speed through the nozzle, which generates the aerosol spray.

Image sources: (Top) Courtesy of Prof. Maria D. King, Texas A&M University.[15] (Bottom) User Wikip2011 on Wikipedia. https://en.wikipedia.org/wiki/Metered-dose_inhaler#/media/ File:Metered-dose_Inhaler.JPG (CC BY-SA 3.0).

And so it is that a spray remains the most common device used for treating asthma and other respiratory conditions. Although conceived for another ailment, this simple droplet-generating mechanism became a routine lifesaver for athletes such as Jackie Joyner-Kersee and millions of other asthma patients, amounting to billions of doses in trillions of droplets yearly.[o] That's microfluidic healing ingeniousness for you.

---

[15] Kesavan, J., Schepers, D. R., Bottiger, J. R., King, M. D., & McFarland, A. R. Aerosolization of bacterial spores with pressurized metered dose inhalers. *Aerosol Sci. Technol.* 47: 1108 (2013).

[o] According to data from the Institute for Mathematical Statistics, in 2014, more than 2,000 people were taking doses from a metered-dose inhaler every second.[14] This amounts to roughly 63 billion doses a year.

Summary

- An *atomizer* or *nebulizer* is a device that creates a microfluidic spray of droplets. Jean Sales-Girons, a French doctor, invented the first atomizer in 1858 to facilitate the delivery of drugs in the form of droplets through the lung's airways for the treatment of respiratory diseases.
- A *metered-dose inhaler* is a modern-day atomizer commonly used to treat asthma.
- Lord Rayleigh first explained in 1878 that when a fluid stream in air breaks up into droplets, it is due to the surface tension of the fluid. This instability of fluid streams in air received the name of *Rayleigh instability* in his honor. The sprays of asthma inhalers and nebulizers are generated by Rayleigh instability.
- A *Venturi* is a fluidic device that exploits the *Venturi effect*, first described by Italian physicist Giovanni Battista Venturi in 1836. The Venturi effect consists of the reduction in fluid pressure that results when a fluid flows through a constriction in a pipe.
- Many spray generators work by blowing air through a constriction that contains a hole, which is connected to a reservoir. Because of the Venturi effect, when air is blown through the pipe, the reduction in pressure at the constriction causes fluid to rush toward the hole. Fluid from the reservoir is ejected into (and collides with) the air current—which generates a spray similar to those generated by Rayleigh instability.
- All existing spray devices are based either on the Venturi effect or on the Rayleigh instability.
- Sprays have had a major impact as vehicles for delivering asthma medication and anesthesia. Their use has been extended to other applications, such as irrigation, insecticide spraying, and cosmetics.

4

Mercedes' Droplets

How Sprays Revolutionized Combustion Engines, from Carburetors to Fuel Injectors

Bertha and Karl Benz (Figure 4.1, top) had been married for sixteen years when Bertha decided to make a bold, and rather scandalous, last-ditch effort to save her husband's languishing business. Careful not to awaken Karl, she quietly got out of bed on the morning of August 5, 1888, and readied herself and her two teenage sons for a surprise trip to her mother's house in Pforzheim, Germany, about 70 miles north of the Benzes' home in Mannheim. Bertha left Karl a note on the kitchen table letting him know where they were going—but, sneakily, not *how* they were getting there.[1]

Bertha and the boys headed to Karl's nearby workshop and stealthily rolled the Benz Patent-Motorwagen III—a snazzy three-seater with three wire-spoked wheels (Figure 4.1, bottom)—out of the garage. This car was hardly ordinary. It was the first production automobile in history, in its third iteration or model. They cranked it up at a safe distance from Karl's bedroom window and were on their way. When Karl saw their note on the kitchen table, he assumed they had taken the train. He did not realize that Bertha had taken the car until he got to his workshop later that morning.

Neither of them could have suspected that, not far from there, a pair of engineers were in the early stages of developing an engine that would have superior performance due to a microfluidic device.

Karl was extremely fortunate to have married Bertha. They had met at a social club outing in 1869, and Karl could not help but be smitten by the fiercely intelligent, wealthy, and beautiful young woman. Bertha saw in Karl a brilliant engineer, and they both shared a fascination with the kind of cutting-edge technological innovations that would result in their first engine-powered horseless carriage. Unfortunately, Karl had neither the means nor the business savvy to finance the automobile factory he dreamed of. When they got engaged, Bertha decided to invest her dowry into what she saw as their shared future, the Benz and Cie automobile company. The transaction was completed before their

---

[1] Tweney, D. Aug. 12, 1888: Road trip! Berta takes the Benz. *Wired.com.* https://www.wired.com/2010/08/0812berta-benz-first-road-trip (2010).

Figure 4.1 (Top left) Bertha around 1870 at 21 years of age. (Top right) Karl Benz in 1869 at 25 years of age. They married in 1872. (Bottom) The Benz Patent-Motorwagen Nr. 3 of 1888, used by Bertha Benz for the first long-distance journey by automobile. The Nr. 1 model, built in 1885 and released in 1886, was similar to this one.

Source: Wikipedia. (Top left) https://en.wikipedia.org/wiki/Bertha_Benz#/media/ File:Berthabenzportrait.jpg. (Top right) https://en.wikipedia.org/wiki/Carl_Benz#/media/ File:Carl_Benz_1869.png. (Bottom) https://en.wikipedia.org/wiki/Carl_Benz#/media/ File:Motorwagen_Serienversion.jpg. Public domain.

marriage in 1872 because, at the time, German law prohibited married women from acting as investors.[2] It was from this shadow seed money that the first Benz Patent Motorwagen was produced in 1886.

Although Karl Benz's skills paid off in the end, he had to compete with several inventors who surpassed him in pursuing the internal combustion engine vehicle. In September 1860, *Scientific American* declared, "The steam age is [sic] ended." French engineer Jean J. Lenoir had invented the Hippomobile, a boxy, four-seater carriage powered by a new "explosion engine"—the first so-called

---

[2] Mercedes-Benz-Group. Bertha Benz. Founders and pioneers. https://group.mercedes-benz. com/company/tradition/founders-pioneers/bertha-benz.html (2022).

internal combustion engine.[a,3] Each flare inside the engine activated the vertical motion of a piston, and a crankshaft[b] transmitted this motion to the wheels very much like in steam engines. The Hippomobile engine—which ran on hydrogen gas instead of liquid fuel—completed a 50-mile course but was relatively inefficient. A walking adult could outpace the Hippomobile at top speed. Although several other engines and car models based on "explosion" were invented within the next few years, none made it to the streets.[c] By insisting on an internal combustion engine design powered by a less volatile but more chemically efficient liquid fuel,[d] Benz became the first to produce and sell a viable, complete automobile.

But those first cars were a hard sell, to say the least. Benz's first model, the Patent-Motorwagen I, was a three-wheeled, two-seater with a 0.75-horsepower (hp) one-cylinder engine and a top speed of 10 kilometers per hour (km/hr), about 6 miles per hour (mph). It cost approximately $1,000 (the equivalent of $30,000 today) to produce. The three-wheeled, three-seater Model III that Bertha took for a ride had 2 hp and a top speed of 16 km/hr (10 mph).[5] By comparison, a standard car today has 120 hp, a sports utility vehicle has up to 200 hp or more, and a flamboyant Ferrari has at least 400 hp. Naturally, horses were the vehicular standard for power and speed in 1888. A horse could reach 50 km/hr (31 mph). At the time, you could purchase a serviceable horse and carriage for about a tenth of the price of the Model III.[6] So Karl and Bertha understood they had a lot of work to do before people would buy their cars.

A key part of the engineering challenge was that engines of the time had primitive, inefficient carburetors. The carburetor is the component of an engine in which fuel becomes mixed with air. All of Benz's early models

[a] The term "internal" was used by opposition to the previous steam engines, which were based on *external* combustion. For a brief introduction to how an internal combustion engine works, complete with excellent three-dimensional animations, see Ciechanowski.[3]

[b] The crankshaft is the part of an engine or machine that translates the vertical motion of the piston (or pistons) into rotation.[3] There is evidence that primitive crankshafts had been invented as early as the third century AD in Hierapolis (Asia Minor) to connect the wheel of a sawmill. Arabic inventor Al-Jazari invented a crankshaft in 1206 to pump water as part of a water-raising machine.[4]

[c] Most notably, the German Nikolaus Otto and his brother Wilhelm were inspired by Lenoir to develop their own design, and they started a company as well. The Otto brothers attracted German engineers Gottlieb Daimler and Wilhelm Maybach, who ended up leaving the company due to disagreements. Otto, Daimler, and Maybach (as well as Benz) are recognized as pioneers of internal combustion engines.

[d] Daimler had recognized the need for liquid fuel at around the same time.

[3] Ciechanowski, B. Internal combustion engine. https://ciechanow.ski/internal-combustion-engine (2021).

[4] Crankshaft. *Wikipedia*. https://en.wikipedia.org/wiki/Crankshaft (n.d.).

[5] August 1888: Bertha Benz takes world's first long-distance trip in an automobile. Daimler. http://media.daimler.com/dcmedia/0-921-1088722-1-1096678-1-0-0-1096723-0-1-11702-61431 8-0-1-0-0-0-0-0.html (2008).

[6] Choosing Voluntary Simplicity. What did things cost in 1872? http://www.choosingvoluntarysimplicity.com/what-did-things-cost-in-1872 (2007).

incorporated a "surface carburetor," essentially a Venturi chamber where air flowed through a constriction to create a low-pressure point (see Chapter 3). The airflow facilitated the evaporation of fuel that then was fed to the engine in gas form.[e] Hence, the surface carburetor required the fuel to be a very volatile solvent. There were a lot of surface carburetor designs, many of them allowing for adjusting the proportion of air that was mixed in with the fuel vapors before entering the engine. Later, a carburetor incorporating microfluidic sprays, which could run on liquid fuels, would increase the mixing efficiency and supersede these designs. However, the morning that Bertha and the boys drove away, the Karl Benz Patent-Motorwagen lacked any boost from microfluidics.

Bertha had done test drives of the first units and, while becoming an adept pilot, she had also developed keen design and marketing instincts. She realized that the money from her dowry was not collecting its fruits. Karl's factory could ruin them if she didn't do anything about it. In August 1888, she decided—without telling her husband but with her children as willing accomplices—that the car business needed an advertising campaign. What could be better than a trip through the countryside to show the car off for everyone to admire her husband's horseless carriage in full daylight?

It all went according to plan, more or less. Bertha and the boys addressed the challenges as they arose. There were no gas stations in 1888. When they realized they were low on fuel, they stopped in Wiesloch at a pharmacy to buy 3 liters of ligroin, a laboratory solvent and paint thinner derived from petrol. The pharmacy still exists today and proudly claims to be the first gas station in history. The insulation of the ignition wire wore through at some point. Still, Bertha—an inventive mechanic—discreetly reached up her skirt and volunteered her garter. She also used her hat pin to clean a blocked fuel line. The shoe brake, which acted on both rear wheels, was hand-operated with a lever on the side of the vehicle. Her fifteen-year-old son Eugen's muscles came in handy here, but the frequent terrain slopes and the vehicle's heavy weight (360 kg) caused the brake shoes to wear out quickly. Bertha stopped at a cobbler to have the brake shoes covered with leather, thus inventing the "replaceable" brake lining right there.[5,f]

Bertha updated Karl by sending him several telegrams along the way. Upon reaching grandma's house in Pforzheim, they had completed more than 100 km,

---

[e] The invention of the surface carburetor design is credited to Italian engineer Enrico Bernardi in 1882.

[f] By modern standards, Bertha Benz would have been listed as a co-inventor in many patents. However, German law at the time did not allow the naming of married women as inventors in patents.

the first long-distance trip by car. It must have been quite a sight for the locals, who had never seen nor heard a horseless, noisy motor car. They would gossip about her at every street corner, shop, and church—which was precisely her plan from the beginning. Bertha succeeded in drawing public attention to her husband's invention.

They needed the publicity because their fiercest competitors—two German engineers named Wilhelm Maybach and Gottlieb Daimler—were just 100 km southeast of Mannheim, near Stuttgart. Maybach and Daimler had been working on a different automobile and engine design, also fueled by ligroin. In 1885, three years before Bertha went for a ride, Maybach and Daimler had finished their "grandfather clock engine" (Figure 4.2), which they later realized was superior to Benz's. The news of Bertha Benz's trip must have been bittersweet to them because their company—the Daimler Motoren Gesellschaft (DMG)—was still not selling cars and would not sell its first car until 1892. But something truly revolutionary happened at DMG in 1893.

That year, Maybach realized that fuel could be more efficiently mixed with air if atomized into a spray of droplets. To produce the spray, he devised a constriction in an air pipe (i.e., a Venturi, see Figure 3.4) that caused sufficient vacuum to eject a jet of fuel from a nozzle, making it collide with the air stream. The air stream atomized the fuel jet into countless droplets, much like a gale blows sea spray from the top of a wave—effectively a very efficient mixture of air and fuel. There is no record of Maybach's source of inspiration, but there are several possibilities.[8] In addition to being familiar with the Venturi principle from working with surface carburetors, Maybach could have noticed the Venturi-based perfume atomizers that were all the rage among fashionable Victorian-era women. Doctors in that era treated respiratory ailments by spraying the medication into patients' throats using Venturi nebulizers (in particular, the Hydrokonium invented in 1862 by a German doctor; see Chapter 3),[8,9] and a medical book from 1884 reviewed a dozen designs with nozzle setups very similar to that of Maybach's.[10]

---

[8] There were a few predecessors to Maybach's invention, although credit has gone mostly to Maybach for various reasons. In 1875, German engineer Siegfried Marcus had come very close to Maybach's realization by building a carburetor that sprayed droplets using rotating steel brushes. In an attempt to rewrite history, during World War II, the Nazi Ministry for Propaganda ordered the removal of Marcus (who had Jewish ancestry) as the inventor of the car from German encyclopedias and instead his name was replaced with the names of Daimler and Benz. And in 1887, British engineer Edward Butler invented a spray carburetor for an engine that powered a tricycle. Although it never became a commercial product, the courts later determined that Butler's spray carburetor predated Maybach's.[7]

[7] Gale, T. Carburetor: Research article from World of Invention. *BookRags*. https://www.bookrags.com/research/carburetor-woi/#gsc.tab=0 (2005).

[8] Scudder, J. M. *On the Use of Medicated Inhalations in the Treatment of Diseases of the Respiratory Organs*. Moore, Wilstach & Baldwin, 1867.

[9] Beatson, G. On spray producers. In *The Glasgow Medical Journal* (ed. Coats, J.), vol. 14. Glasgow and West of Scotland Medical Association, 1880.

[10] Prosser, J. *The Therapeutics of the Respiratory Passages*. William Wood, 1884.

Figure 4.2 The grandfather clock engine by Maybach and Daimler, still incorporating a surface carburetor (the cylindrical chamber at center top). Later models would incorporate carburetors based on spray nozzles with a small modification.

Image source: Wikipedia user Morio. https://en.wikipedia.org/wiki/Gottlieb_Daimler#/media/ File:Daimler_Standuhr_engine_2_Mercedes-Benz_Museum.jpg (CC BY-SA 3.0).

Whatever his inspiration, Maybach had just devised a spray-nozzle carburetor, which he fitted into the new two-cylinder "Phoenix" engines.[11] The ignition of the spray inside the cylinder expanded rapidly to move a piston, and the

---

[11] Mercedes-Benz-Group. 1886–1920. Beginnings of the automobile. https://group.mercedes-benz.com/company/tradition/company-history (2022).

crankshaft transmitted this motion to the wheels.[3] After Maybach's design, Benz invented the carburetor "butterfly" valve (now called "throttle valve," used in the carburetors of generations to come; see Figure 4.3), which connected the carburetor and the engine and allowed for regulating the fuel–air mixture, facilitating the adjustment of thrust.[7] For a modern four-cylinder engine running at a modest speed of 1,500 revolutions per minute, the crankshaft turns at 25 revolutions per second. Each cylinder fires once per two revolutions of the crankshaft, but because there are four cylinders, there are approximately 50 explosions per second in the engine occurring at that speed. The goal of each of these explosions is to achieve the combustion of all the fuel droplets (to produce the maximum torque), although that's not always achieved at 100% efficiency. Nevertheless, bringing droplets into engines was a revolutionary idea that changed the history of the automotive industry.

While DMG was catering to customers who wanted to ride fast,[h] Karl Benz aimed for affordability and ease of handling rather than power at the urging of

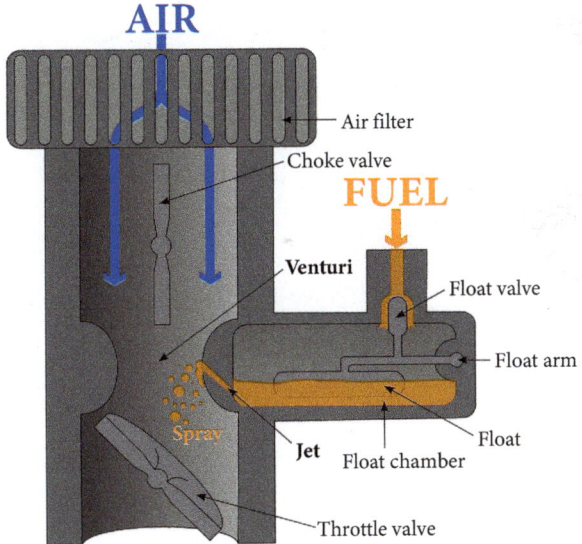

Figure 4.3 Cross-sectional schematic of a downdraft carburetor, featuring a vertical air duct and a horizontal fuel chamber. This design differs slightly from the original one by Maybach, but they are both based on the Venturi effect.

Image source: User K. Aainsqatsi on Wikipedia. https://en.wikipedia.org/wiki/Carburetor#/media/ File:Carburetor.svg (CC BY-SA 2.5).

---

[h] DMG incorporated Maybach's novel carburetor into engines that produced 2- to 6-hp output to propel the four-seat, four-wheel DMG Riemenwagen models.[12] The fastest, 6-hp model could reach 25 km/hr.

[12] Mercedes-Benz-Group. Daimler belt-driven car, 1895–1899. https://mercedes-benz-archive. com/marsClassic/en/instance/ko/Daimler-belt-driven-car-1895-1899.xhtml?oid=5929 (2022).

his board.[i] By 1898, Karl Benz had designed the "Velo" (short for Velociped), a simple and light vehicle with 1.5–3 hp (reaching 20–30 km/hr) and a price of 2,200 Marks. The Velo became the first mass-produced car.[j] "This vehicle was quite literally grabbed out of our hands," Benz said in 1909. "What we made was sold immediately."[13] At the very least, Bertha's trip demonstrated to the blossoming automotive industry that test drives were essential to their business. It would also be difficult to imagine the rise of the later Benz & Cie. Rheinische Gasmotorenfabrik AG—which became the biggest automotive factory in the world—without Bertha's publicity stunt.

These developments excited not just engineers. On July 22, 1894, the Parisian newspaper *Le Petit Journal* organized the first motor car race in history, the 126-km route from Paris to Rouen, which was massively attended. The race was not just about speed, as the horseless vehicles had to fulfill the criteria of safety, ease of handling, and low operating cost, and the participants stopped a couple of times along the way to exchange impressions and have lunch. Overall, the enthusiastic conclusion was that combustion engines were far superior to all other forms of propulsion, including horses and steam. "How can you travel other than in a motor car?" summed up *Le Petit Journal* the following day. Of 21 cars, the two vehicles that ended first were DMG Riemenwagen with a Phoenix engine. Benz's Velo arrived fifth.[14]

Car racers took notice of the Phoenix engines. In March 1899, wealthy car enthusiast Emil Jellinek entered a DMG 12-hp Phoenix-engine model in the Nice–Magagnone–Nice race under the pseudonym of "Mercedes," his ten-year-old daughter's name. DMG factory driver Wilhelm Bauer won the race. A year later, Jellinek tried to repeat the victory with a 23-hp model. Still, the engine was too powerful for the car's design, and Bauer was killed in a crash during the race. Jellinek then demanded that DMG completely redesign the vehicle with a lower center of gravity, less weight, and easier handling. The result was the 1901 35-hp Mercedes, featuring a four-cylinder engine[3] and a spray-nozzle carburetor—the first car to reach 75 km/hr (Figure 4.4). It won many races and was an instant success despite its high price of 16,000 Marks. "Nous

---

[i] In 1893, Benz invented the mechanism called double-pivot steering that solved the problem of steering four-wheeled vehicles.[11] After the Motorwagen Nr. 3, from 1893 until 1900 he built the Victoria, a single-cylinder, four-seater car with four wheels that could output up to 6 hp and reach 30 km/hr, at a cost of between 4,000 and 5,000 marks.

[j] Here, "mass-produced" means that it was produced in series. The first automobile produced in an assembly line was the 1901 Curved Dash Oldsmobile, which sold for $650. Approximately 19,000 units were fabricated between 1901 and 1907.

[13] Tahaney, E. The Benz Velocipede Ruled the Automotive World 125 Years Ago. *Motortrend.com.* https://www.motortrend.com/vehicle-genres/1894-benz-motor-velocipede-first-mass-produced-car/ (2019).

[14] Mercedes-Benz-Group. Mercedes 35 hp racing and touring car, 1900/1901. https://mercedes-benz-publicarchive.com/marsClassic/en/instance/ko/Mercedes-35-hp-racing-and-touring-car-1900--1901.xhtml?oid=6729 (2022).

Figure 4.4 The 1901 Mercedes model, which was much faster and more powerful than its contemporaries. Like its predecessor the Phoenix, it already incorporated Maybach's microfluidic spray-nozzle carburetor.

Image source: © Mercedes-Benz Group AG. https://mercedes-benz-publicarchive.com/marsClassic/ en/instance/ko/Mercedes-35-hp.xhtml?oid=5901.

sommes entrés dans l'ère Mercédès" ("We have entered the Mercedes era"), wrote the secretary general of the French automobile club in a review that year.[15]

The cars of Benz and Daimler (who never met) did not compete much with each other in the end. Benz quickly adopted spray-nozzle carburetors, and in any case, Benz's cars did not need the power because they were targeting the broader public and roads were not even paved back then. In 1926, as soon as the German economic crisis hit, Benz and DMG—by then the two oldest automobile manufacturers in the world—agreed to merge as Daimler-Benz.

Other inventors realized there were other ways to create sprays of fuel droplets. *Fuel injectors* were one of them. Initially, fuel injectors were not competitive enough, so carburetors were used until the late 1980s. However, carburetors had an Achilles' heel: It was difficult to tune the spray size with an air stream. Thus, in practice, a carburetor-based engine did not work efficiently when the car was idle—a significant disadvantage as fuel prices began to rise. Carburetors wasted a lot of gas, polluting unnecessarily and generating deposits that had to be cleaned periodically. The mechanically simpler pneumatic fuel injector solved these problems (Figure 4.5).

The function of a fuel injector is essentially the same as that of a carburetor: to atomize the fuel into millions of droplets through a nozzle so the fuel

[15] Mercedes-Benz-Group. Daimler "Phoenix" car, 1897–1902. https://mercedes-benz-publicarch ive.com/marsClassic/en/instance/ko/Daimler-Phoenix-car-1897-1902.xhtml?oid=5982 (2022).

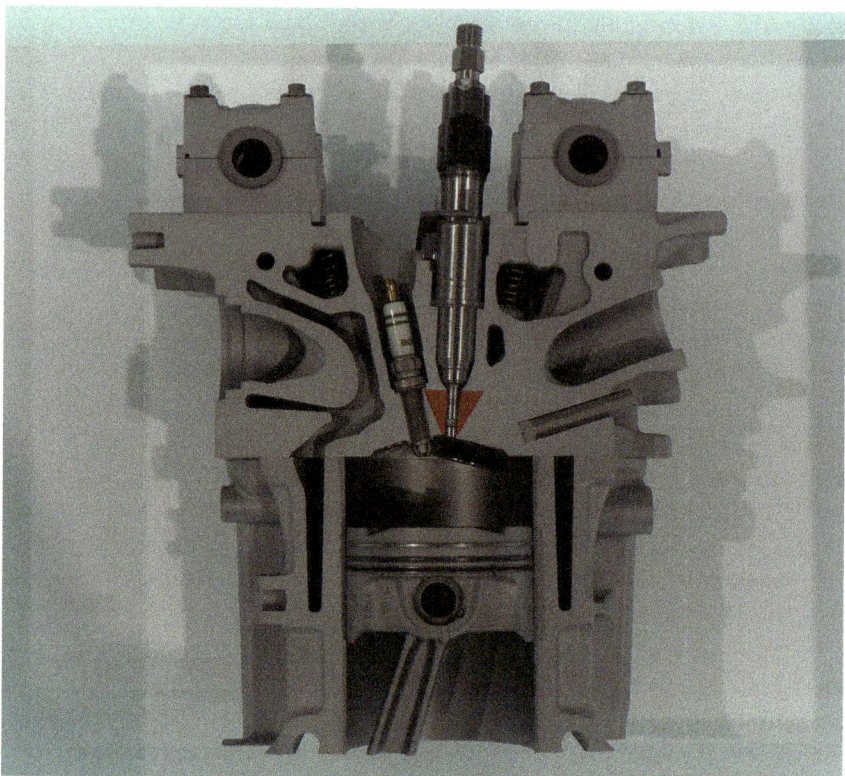

Figure 4.5 A cutaway model of a modern internal combustion engine with fuel injection. Injection occurs in the form of a downwards spray at the tip of the red arrow (the nozzle). As the fuel droplets enter the chamber just above the cylinder and to the right of the spark plug, the spark plug is electronically coordinated to ignite the fuel. This explosion pushes the cylinder down, transmitting torque to the crankshaft.

Image source: User Ton1~commonswiki on Wikipedia. https://en.wikipedia.org/wiki/Fuel_injection#/media/File:PetrolDirectInjectionBMW.JPG (CC BY-SA 3.0).

can be mixed with air before ignition. Fuel injectors are more precise and consistent because they are calibrated to force the same amount of fuel into each cylinder. (In carburetors, on the other hand, the mixture of air and fuel is produced inside the carburetor, from which it enters the cylinders.) Instead of using the suction created by the Venturi effect, a fuel injector pumps the fuel at very high pressure through the nozzle, creating a fuel jet. As soon as the jet contacts (still) air, surface tension slices the unstable jet into a spray of droplets (recall the Rayleigh instability from Chapter 3), with a cone angle that depends on the

jet speed.[16] Before ignition, the droplets in the spray travel at more than 60 meters per second for a few thousandths of a second.

Most people are surprised to learn that fuel injectors predated the carburetor before taking several decades to catch up to it. The first patent on an internal combustion engine equipped with a fuel-injection system was granted to American engineer George Bailey Brayton[17] in 1872, more than fifteen years before Bertha Benz took the boys for a ride and more than twenty years before Maybach came up with his spray-nozzle carburetor. Brayton's engine was never incorporated into a car. Still, German inventor Rudolf Diesel copied Brayton's design and improved the injector—in his 1895 patent,[18] he wrote of "finely divided fuel" and a system of compression now known as "air-blast injection"—although it did not perform very well initially. Improvements in the injection system over a decade led to the nozzle design being used on all diesel engines from the beginning of the twentieth century onwards. The Wright brothers included one in their first airplanes around that time. But these first fuel injectors were not electronic and were less efficient than carburetors. Hence, our parents and grandparents inhaled the fumes of carburetors or diesel engines for several generations before the advent of electronic fuel injectors.

In the modern electronic fuel injection systems, introduced in the 1970s and a standard since the early 1990s, the ratio of fuel and air can be more accurately controlled than with carburetors. An electronic fuel injection system has sensors at the air intake and exhaust that can adapt the ratios in real time based on the throttle, the air temperature, and the exhaust gases. All car manufacturers now use electronic fuel injectors to comply with emission requirements and consumer demands on fuel efficiency. Here is an homage to mechanical minimalism.

Nowadays, fuel injection systems support virtually every non-electric engine for land, sea, and air transportation. In 2020 alone, 40 million flights carried more than 4 billion air travelers. Every year, boats transport more than 11 billion tons of cargo. Nearly 1.5 billion four-wheeled vehicles circulate globally, including 1 billion passenger cars. Add to that almost 300 million motorcycles. The point is that we would not be moving much if it were not for these microfluidic sprays that have stayed hidden under the hood since 1888, when Bertha and the boys took the Benz Patent-Motorwagen III to go visit grandma. For a century and a half, every time a car purred, a truck chugged, a moped throttled, a boat

---

[16] Lefebvre, A. H., & McDonell, V. G. *Atomization and Sprays.* CRC Press, 2017.

[17] Brayton, G. B. Improvement in gas-engines. https://patents.google.com/patent/US125166A/en (1872).

[18] Diesel, R. Internal-combustion engine. https://patents.google.com/patent/US608845A/en (1898).

rolled, or a plane flew, tiny fuel droplets powered blast by blast a monumental, dizzying movement of people and goods throughout the world.

Summary

- The efficiency of early combustion engines depended much on the carburetor, the component of a combustion engine in which liquid fuel is mixed with air.
- The *spray-nozzle carburetor* is an early type of carburetor invented by German engineer Wilhelm Maybach in 1893 that generated a spray of fuel droplets using the Venturi effect.
- The *fuel injector* is another device for mixing liquid fuel with air that also generates a spray of fuel droplets, but using the Rayleigh instability instead of the Venturi effect.
- Fuel injectors did not displace carburetors until the much more precise, computer-controlled *electronic fuel injectors* became the standard in the early 1990s.
- With both carburetors and fuel injectors, sprays of droplets have fueled the ignition of billions of combustion engines since the very first days of the automotive industry.

5

This Olive Is Not an Olive

Similarities in the Use of Droplets in Gastronomy and Genomics

Despite never attending college, the Barcelona-born, three-Michelin-star-chef Ferran Adrià was so highly regarded in academic microfluidic circles that Harvard University invited him to give a series of wildly popular lectures in 2010. Adrià not only *is* a genius but also has the *air* of a stereotypical one, with his frizzy gray hair crowning a broad forehead and thick eyebrows intensifying his fiery, curious eyes. At the age of 18, he quit school and started working in various restaurants to pursue his goal of becoming a chef. Four years later, in 1984, he was hired by El Bulli, a small restaurant perched on Cala Montjoi, a rocky cove at the end of a winding, narrow road in the Costa Brava, two hours north of Barcelona. He was promoted to head chef a year and a half later. The restaurant meteorically rose to fame under the explosive ingeniousness of Adrià. *Restaurant* magazine recognized El Bulli as the best restaurant in the world a record five times between 2002 and 2009. In 2004, *TIME* magazine chose Adrià as one of the 100 most influential people alive.

At El Bulli, guests could taste his "deconstructivist" culinary creations, designed to "provide unexpected contrasts of flavor, temperature, and texture. His motto, reminiscent of René Magritte's painting *Ceci n'est pas une pipe* ("This Is Not a Pipe"), was "Nothing is what it seems." "The idea is to provoke, surprise, and delight the diner," he said.[1] Top chefs widely agree that Adrià revolutionized modern gastronomy in the 1980s with the invention of molecular gastronomy. At the same time, he mesmerized the guests of El Bulli with unimaginable textures and novel presentations, such as edible "foams" that are generated by injection of carbonation in a sauce with a soda siphon. Several of his signature plates contained "spherified" foods (Figure 5.1). Widely used in biotechnology and microfluidics, spherification is a process that uses calcium ions to trigger the

---

[1] Adrià, F., Soler, J., & Adrià, A. *El Bulli 1983–1993*. RBA Practica, 2004.

Figure 5.1 Spherified olives and peanuts at El Bulli. These edible constructs taste like olives and peanuts but are made from oil droplets, yielding a gelatinous, almost liquid-like, texture that confuses the palate in a creative way. Adrià used various "molecular gastronomy" approaches to "deconstruct" the flavor from the texture, form, and temperature in order to dazzle his guests.
Image by Albert Folch.

gelation of alginate, an edible gel derived from seaweed.[a] For example, two of these gelatinous constructs tasted like olives or peanuts but had an unexpected liquid-like texture that purposely startled the senses.

In addition to his spherified foods, many of Adrià's activities captured the attention of microfluidic engineers. El Bulli closed in 2011, but to stay creative, Adrià started elBullifoundation, a set of research and educational projects around the science of cooking, including a cookery school and an encyclopedia of curated recipes called the *Bullipedia*.[2] The entry for mayonnaise, for example, describes the scientific process of making an *emulsion* from egg yolk, oil, and vinegar. "In the end, everything is science, everything," he says[2]—a remarkably profound statement coming from someone without a scientific education.

If you look under the microscope, you will see that all sauces, from mayonnaise to gravy, are formed of droplets. Unlike the mid-air droplets created by metered dose inhalers or carburetors, these droplets are surrounded by another

---

[a] The spherification process was first patented in Britain in 1942 by William Peschardt, a food scientist working for the firm Unilever, and demonstrated by the representatives of a Spanish food company to Adrià, who then made it famous starting around 2003.[3] The process entails thoroughly mixing your liquid food of choice—say, an olive purée or a fruit punch—with alginate powder and then dropping it into the calcium-containing solution. Alginate stays liquid in the absence of calcium and polymerizes into a gel in the presence of calcium. After a few minutes, the outer layer of the alginate mixture gels, creating a soft liquid "bubble" that can be safely manipulated but easily pops in your mouth for your enjoyment.

[2] Williams, G. After elBulli: Ferran Adrià on his desire to bring innovation to all. *Wired.* https://www.wired.co.uk/article/staying-creative-ferran-adria (2012).

[3] ChefSteps. The science of spherification. https://www.chefsteps.com/activities/the-science-of-spherification (2024).

Figure 5.2 Water droplets containing food coloring dyes in olive oil forming an emulsion.
Courtesy of Andrew deMello, ETH Zurich.

fluid. Something magic occurs at the microscopic interface between the two fluids because they resist mixing. Yet, a spoonful of them tastes and feels like they have mixed. These unique "microfluidically-separated-mixtures" of oil and water are called *emulsions* (Figure 5.2). Both oil droplets surrounded by water and water droplets surrounded by oil are called emulsions.

Emulsions were likely invented millennia ago in the context of cooking. Still, they are not the monopoly of chefs and cooks. The Romans made emulsions with beeswax, olive oil, and rose water, which they used as skin moisturizers, not unlike the ones found in pharmacies today. Pharmaceutical companies and pharmacists often concoct drugs that are soluble in oil but not in water. To dilute a drug to optimal effect and make it more absorbable through the

skin, they mix it with water in creams or ointments, which are also emulsions. People buy cosmetic pastes or lotions to revitalize or exfoliate their skin and shave unwanted hair by applying shaving cream; all these products are emulsions. Whenever you apply shampoo, conditioner, moisturizer, or gel to your mane, you spread an emulsion of various products designed to lower the surface tension of water, such as soap, together with an oily fragrance. Here, the emulsion also serves the additional purpose of diluting the strong effects of the smell. These odorants bind to olfactory receptors in our nose; at very high concentrations, the saturation of the receptors often produces a pungent smell sensation, whereas at a more dilute concentration, the smell is floral and sweet. While diluting a water-soluble substance is trivial, most odorants are oils. Thus, chemists make emulsions of fragrant oils in water to dilute them and use them as perfumes. Chemists also make emulsions to synthesize substances such as glue or paint. Everyone has been using these droplet mixtures for quite some time.

As the saying goes, water and oil don't mix. Still, you can vigorously whisk a water-based solution such as vinegar and oil to attempt to form salad dressing. However, by the time your emulsion makes it to the table, you will notice that the oil and vinegar have separated. What happened? The emulsion was unstable because as the vinegar droplets collided, nothing prevented them from joining or "coalescing." Chemists and chefs use molecules called *emulsifiers* to stabilize an emulsion. For your vinaigrette, you could try adding mustard or mayonnaise. Both contain lecithin, a natural emulsifier found in egg yolks and soybeans and in lower concentrations in mustard seeds. Emulsifiers are surfactants, molecules that insert themselves at the surface of a fluid, reducing the surface tension.[b] Emulsifiers coat the droplets within an emulsion and prevent them from merging. Lecithin is an *amphiphilic* molecule, which means it binds to water on one side and to oil on the other, acting as a stable bridge for the two. The fat-loving end of lecithin is soluble in oil and inserts itself into the oil droplets, preventing them from merging with other droplets. In contrast, the water-loving end of lecithin is soluble in vinegar, which makes the droplets stable in vinegar.

We find this mechanism in many sauces. Mayonnaise is an emulsion of oil and vinegar that uses lecithin-containing egg yolk as the emulsifier. Hollandaise sauce is an emulsion of butter and lemon (Figure 5.3). Béarnaise sauce is an emulsion of clarified butter and white wine vinegar. Both sauces use egg yolk as the emulsifier. The emulsifying properties of garlic led to the sauce called aioli (after the Catalan "allioli," which means "garlic and oil," its only ingredients).

---

[b] You have seen the emulsification mechanism in action when you wash a grease stain away with soap. All detergents are surfactant molecules with a fat-loving end and a water-loving end. When grease gets exposed to detergent, the fat-loving ends insert themselves into the fat and coat it with the water-loving ends—which helps the grease dissolve in water. Next time you shower, or wash your dishes or clothes, thank microfluidic droplets for the marvelous experience of soap's cleansing action.

Figure 5.3 Sauces are emulsions. Depicted is the emulsification of eggs in butter and lemon to produce hollandaise sauce.
Image source: User Maria Polna on Pxhere.com. https://pxhere.com/en/photo/1618861.

Although solid, chocolate starts as an emulsion of milk and cocoa butter. Milk is an emulsion, too—a complex emulsion of water, protein particles, and fat. In all cases, the water helps you spread the flavor in the fat droplets across your mouth. You have been drinking droplets since the day you were born.

* * *

Just like the laws of electricity are the same for both positive and negative charges, the principles of wetting and surface tension apply both to droplets and to those voids surrounded by a liquid we call bubbles. You enjoy microfluidic bubbles daily when you ingest bread; a cake; a croissant; a muffin; or a sparkling beverage such as soda, beer, or champagne. The baker puts yeast in the mother dough to make it rise, and the brewer puts yeast in the wort, the sweet starter drink for beer. These are a type of foam. Without yeast, bread would be as dense as pasta, and beer would be a flat, sugary mix.

Yeast is a single-celled microorganism related to mushrooms. Instead of using sunlight for energy, which virtually all plants do, yeast feeds on sugars such as those that are abundant in fruits, producing alcohol as a by-product—a process called *fermentation*. Yeast is so light that it floats in the air. It can be taken by the wind, so it is not surprising that the fermentation of wine and beer was discovered—probably accidentally from the fermentation of fruit juices—several thousand years ago by several cultures. Stand under a fruit tree and microscopic yeast cells will rain down on you. Approximately 1,500 yeast species exist, but cooks and brewers use mostly one species, *Saccharomyces cerevisiae* ("sugar-eating fungus of the beer" in Latin). As yeast feasts on the sugars in the dough or beer, it produces carbon dioxide as part of its metabolism—as our cells do.

Figure 5.4 The spongy texture of bread is the result of microbubbles formed in the dough by yeast during the fermentation process.
Image source: Pxhere.com. https://pxhere.com/en/photo/589221.

The fluffiness of baked dough (Figure 5.4) originates from the carbon dioxide microbubbles that yeast produces during fermentation. The baker lets the dough rise—that's when the yeast grows and reproduces—and then puts it in the oven to bake it. The heat of baking expands the gas bubbles, causes the flour proteins (gluten) to trap the expanded gas bubbles irreversibly, and develops a delicious flavor from chemical changes to the flour and the now-dead yeast proteins. Look at bread under a microscope at high magnification and you will see the dough and thousands of tiny pores tunneling through the dough (Figure 5.5). Now soak this bread in your favorite sauce, watch the sauce wick into the pores, and enjoy what everyday-capillarity does for you.

Unlike the foams artificially generated with a soda siphon by Adrià, the foam atop your beer (Figure 5.6) is due to carbon dioxide bubbles generated by yeast during fermentation, just as it happens in bread making. Brewers can make beer as top-fermenting and bottom-fermenting, depending on where the yeast is during the process. *Saccharomyces cerevisiae* rises to the top of the brew during fermentation and is used to brew the pale ale, ale, and stout varieties of beers. On the other hand, the yeast *Saccharomyces pastorianus* ("sugar-eating fungus of the shepherd") settles at the bottom during the fermentation of lager and pilsner beers. Every time you munch on a soft pastry or a baguette sandwich, every occasion you sip a pint with friends or a glass of sparkling wine at a cocktail party, the fun relies on microfluidics.

In the lab, bubbles are the enemy of most microfluidic engineers because they interfere with the proper flow of fluids. Like their cousins, the droplets, bubbles hold their shape due to surface tension. To prevent the formation of bubbles inside a chip during regular operation, it helps to coat the surfaces with a water-friendly material, causing any small-nucleated bubble to roll away as soon as it

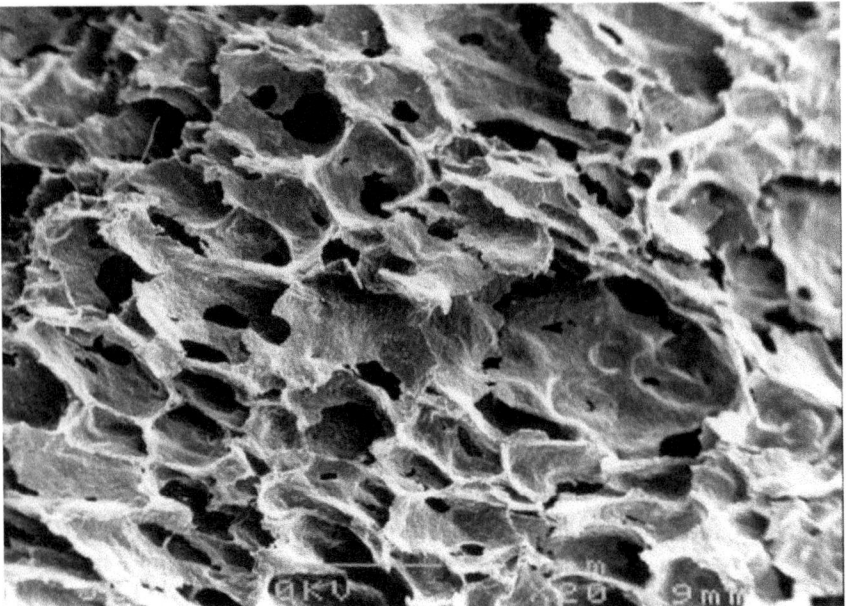

Figure 5.5 Scanning electron micrograph of a piece of torn white bread. The whole image is approximately 6 millimeters wide. As the bread is dipped in a sauce, the liquid wets the walls of the pores and surface tension wicks the sauce into each cavity.

Courtesy of Prof. Claire Davis and Chris Hardy, School of Metallurgy and Materials, University of Birmingham. https://www.flickr.com/photos/core-materials/4419088363 (CC BY 2.0 Deed).

starts forming. Although a gas such as air appears indeed separated from water in a microbubble, gases can also intermingle with fluids by dissolving in them, like the salt we find dissolved in seawater. If you bring a glass of cold water to a warm room and let it sit undisturbed (e.g., the glass of water by your bed in your heated bedroom in the winter), you will notice the appearance of bubbles on the walls of the glass after the glass has warmed up. (It takes a few hours.) That happens because the solubility of gases in fluids depends strongly on the temperature. When the air molecules dissolved in water become insoluble, they cling onto the closest surface they can find and preferentially join other air molecules—that's when you see a full-grown bubble.

Dave Weitz, Harvard professor of physics and a world expert in microfluidics, bridged the study of microfluidics in the lab with microfluidics in the kitchen. Perhaps drawn to Adrià's cooking like a bubble to another bubble or a droplet to another droplet, he invited Adrià to Harvard in December 2008 to talk about science and cooking. "The attraction I had originally to Ferran is that he thinks like a scientist," noted Weitz[4]—an extraordinary compliment given that Adrià

[4] Folch, A. *Hidden in Plain Sight: The History, Science, and Engineering of Microfluidic Technology.* MIT Press, 2022, p. 60, 61 and 63.

Figure 5.6 The frothy foam on beer is made of carbon dioxide microbubbles produced by yeast during the brewing process.
Image source: *Pxhere.com. https://pxhere.com/en/photo/1580847.*

has no science education beyond high school. The talk, attended by several hundred people, was an unprecedented hit, with huge lines forming out the door.[5] Adrià demonstrated melon "caviar" and "pasta" made of ham as examples of his deconstructivist cuisine.[6] Because of the success of that first talk, in 2010, Weitz and other Harvard instructors started the popular class "Science and Cooking,"[7] which is still taught. "It's a way to convince people that it's fun and that there is a lot of stuff we understand from a scientific point of view that chefs exploit," said Weitz, who co-teaches the course.[6]

Weitz has spent his career perfecting microfluidic devices that manipulate droplets at ever-increasing rates. To generate droplets, two oil streams pinch one stream of water into a small channel, forcing the two immiscible fluids to "take turns" entering the small channel and forming thousands of droplets per second (Figure 5.7). Weitz's microfluidic devices allow for studying how an emulsion

[5] Beans, C. Science and culture: Universities move science labs to the kitchen. *Proc. Natl. Acad. Sci. USA* 117(35): 20982–20985 (2020).

[6] Black, J. Foam 101? Chefs Andrés, Adrià will teach at Harvard. *The Washington Post* (2010).

[7] Sweeney, S. In good taste. *The Harvard Gazette.* https://news.harvard.edu/gazette/story/2010/09/in-good-taste (September 8, 2010).

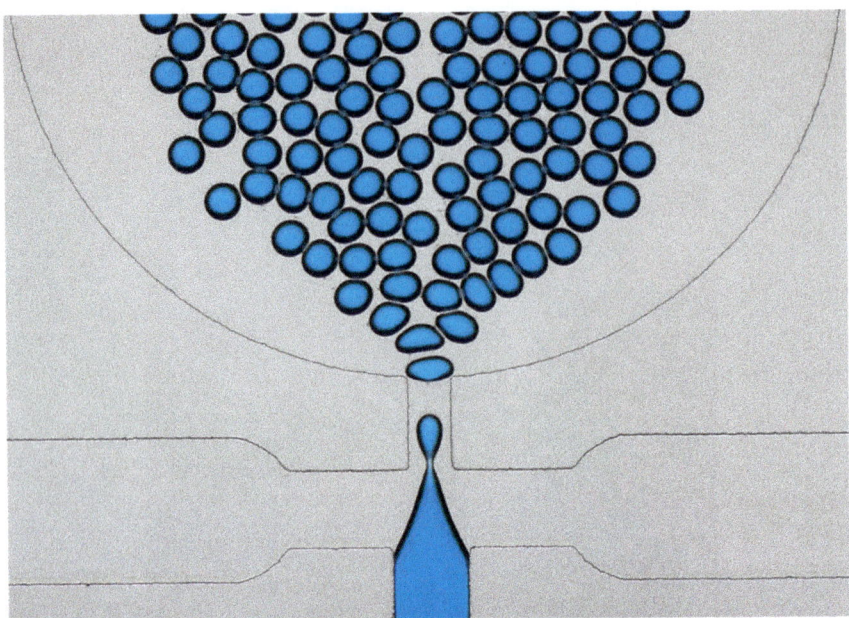

Figure 5.7 Snapshot of the formation of an emulsion of blue droplets in a microfluidic device. A stream of blue dye entering from the bottom becomes pinched from left and right by two streams of transparent mineral oil and emerges through a 100-micrometer-wide channel, generating 80-micrometer-diameter blue droplets surrounded by oil at a rate of 240 droplets per second.
Image courtesy of The Lutetium Project/Microfactory.

forms in real time, one droplet at a time. The devices can also split, mix, and analyze the droplets one by one downstream at dizzying speeds. A large team led by Weitz, Andrew Griffith, and Christoph Merten showed that keeping cells—and even worms—alive inside droplets for several days was possible.[8,9] Chemist Andrew deMello, another microfluidic droplet pioneer, also supports doing big chemistry on a small scale. He said,

> Chemists have for over two hundred years used test tubes and flasks for chemistry only because of ergonomics. A test tube fits in your hands; you can shake it and look into it. But there is no reason why you do that apart from convenience. . . . When you think about simple scaling laws, things should work better, more quickly, and in a more controlled manner when you shrink them down.[4]

[8] Clausell-Tormos, J., Lieber, D., Baret, J.-C., El-Harrak, A., Miller, O. J., Frenz, L., et al. Droplet-based microfluidic platforms for the encapsulation and screening of mammalian cells and multicellular organisms. *Chem. Biol.* 15: 427–437 (2008).

[9] Köster, S., Angilè, F. E., Duan, H., Agresti, J. J., Wintner, A., Schmitz, C., et al. Drop-based microfluidic devices for encapsulation of single cells. *Lab Chip* 8: 1110 (2008).

Some of Weitz's devices use the spherification process in their microfluidic devices. Oil can be toxic to cells in the long run, so spherification allows researchers to package molecules and cells into droplets not surrounded by oil.[10-13] In their devices, the oil is only used briefly and then discarded. The idea is deceivingly simple and starts with devices containing oddly symmetrical features that could have been conceived by someone like the impressionist painter Paul Klee (Figure 5.8). Stefanie Utech, the student who led the study in the Weitz lab, designed the shapes in the channels to ensure that the two oil streams pinched the droplets at vertiginous speed and Swiss clockwork exactitude.

To prepare her experiment, Utech first had to fill the device with fluids and remove the occasional bubble that formed inside. The researchers used the microchannel device to repeat an alginate reaction triggered by calcium ions millions of times (for every droplet). Initially, the calcium ions were sequestered into another molecule by the lengthy name of ethylenediaminetetraacetic acid (EDTA) that keeps the calcium ions tight as long as the environment is not acidic. A mixture of the cells, alginate, and EDTA in a neutral environment was introduced through the top inlet (see Figure 5.8). The oil was introduced in the middle inlet and contained a small quantity of acetic acid. Hence, when the alginate fluid met the acidic oil, droplets formed, and at the same time, the EDTA molecules felt the acidic environment, triggering the release of calcium ions and the spherification of alginate. The spherification reaction occurred for every droplet and took no more than a few thousandths of a second—more than twenty times faster than the blink of an eye. The researchers collected

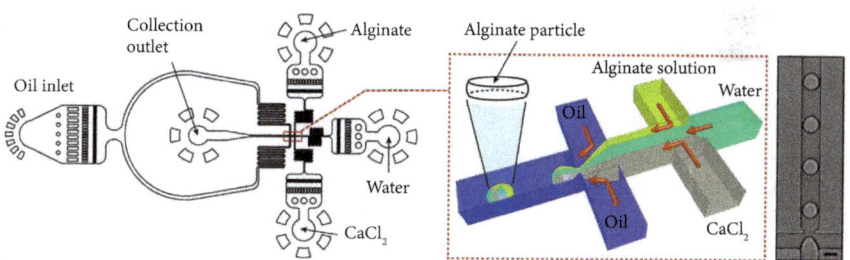

Figure 5.8 Microfluidic device for generating an emulsion of cell-containing alginate drops using the spherification process.
Adapted with permission from a figure by David Weitz,[11] Harvard University.

[10] Tan, W. H., & Takeuchi, S. Monodisperse alginate hydrogel microbeads for cell encapsulation. *Adv. Mater.* 19: 2696–2701 (2007).

[11] Mazutis, L., Vasiliauskas, R., & Weitz, D. A. Microfluidic production of alginate hydrogel particles for antibody encapsulation and release. *Macromol. Biosci.* 15: 1641–1646 (2015).

[12] Utech, S., Prodanovic, R., Mao, A. S., Ostafe, R., Mooney, D. J., & Weitz, D. A. Microfluidic generation of monodisperse, structurally homogeneous alginate microgels for cell encapsulation and 3D cell culture. *Adv. Healthc. Mater.* 4: 1628–1633 (2015).

[13] Chen, Q., Utech, S., Chen, D., Prodanovic, R., Lin, J. M., & Weitz, D. A. Controlled assembly of heterotypic cells in a core-shell scaffold: Organ in a droplet. *Lab Chip* 16: 1346–1349 (2016).

an emulsion of homogeneous, cell-containing alginate droplets in oil through the bottom outlet. But, unlike most emulsions requiring an emulsifier to prevent the droplets from coalescing, the emulsifier was not needed. In fact, they could remove the oil at the end of the process. Spherification was appealing to the palates of El Bulli's visitors because it packaged foods into fun bubbles that exploded inside their mouths—but for microfluidic engineers, the opposite is true: Spherification is a tool for the oil-free, safe packaging of molecules, cells, and tissues.

Like many others, Weitz has been interested for a long time in the genetic analysis of single cells. All the billions of brain cells, cardiac cells, gut cells, and all the other cell types in the body derive from a single cell during development in the embryo. They all carry the same DNA instructions—they turn on part of the genes and turn off the rest, each cell deciding which ones to turn on or off as a function of developmental time, their position in the body, and in reaction to environmental conditions such as a wound or a disease like cancer. Figuring out this cellular "atlas" of the human body has been the holy grail of many researchers for quite a while. Still, it was not attainable before the advent of high-throughput genetic analysis (see Chapter III.F). Weitz collaborated with a large team at Harvard led by Steve McCarroll, who developed a breakthrough based on analyzing large numbers of cells using droplets. Into each droplet, they introduced one cell, a unique bead carrying a sort of "identity tag" for the cell and all the reactants required for DNA sequencing simultaneously.[14] It's easier said than done, but it suffices to say that this level of precision is possible. The bead was "barcoded" with unique DNA molecules serving as the cellular tags, allowing for tracking the cell's identity during genetic sequencing.

To demonstrate the technique—called *Drop-Seq*, short for "droplet sequencing"—the team processed approximately 45,000 cells from the retina through the microfluidic device, identified all ten expected retinal cells for which molecular identifiers exist, and—as a bonus—was able to distinguish as many as thirty-nine subtypes of cells. This technique opened many avenues. In 2017, Adam Abate's group at the University of California, San Francisco isolated, fragmented, and barcoded the genomes of more than 50,000 bacteria with droplets in just one experiment for parallel sequencing.[15]

Similar experiments with tumor cells in droplets enable the identification of cancer-related mutations in thousands of individual cells. Microfluidic droplets

[14] Macosko, E. Z., Basu, A., Satija, R., Nemesh, J., Shekhar, K., Goldman, M., et al. Highly parallel genome-wide expression profiling of individual cells using nanoliter droplets. *Cell* 161: 1202–1214 (2015).

[15] Lan, F., Demaree, B., Ahmed, N., & Abate, A. R. Single-cell genome sequencing at ultra-high-throughput with microfluidic droplet barcoding. *Nat. Biotechnol.* 35: 640–646 (2017).

are helping biologists tame the immensity of the numbers of biology. Ferran Adrià was right: Everything is science in the end. Droplets do not make distinctions and use the same laws of physics and chemistry to help chefs and scientists alike.

Summary

- An *emulsion* is the name given by chemists to the stabilized suspensions of microfluidic droplets of oil in water, or vice versa.
- We eat droplets every day. Examples of emulsions are many edible sauces, dressings, and milk.
- The absence of a droplet—a bubble—is a microfluidic entity that is exploited during the process of *fermentation* to make various doughs (e.g., in breads and cakes), foams (e.g. in beers), and sparkling beverages (e.g., champagne).
- Droplet-generation methods have made it into *haute cuisine* by the hand of the world-renowned chef Ferran Adrià, even though he lacked a scientific background.
- Many cosmetic products (e.g., moisturizers, ointments, shampoo, gel, and conditioner), paints, and adhesives are also emulsions. We use droplets every day.
- To generate droplets containing live cells in microfluidic devices for biotechnological applications, researchers use *spherification*—a process that uses calcium ions to trigger the gelation of *alginate*, an edible gel derived from seaweed.
- By enabling thousands of chemical reactions in parallel, droplet microfluidics has also revolutionized the genetic analysis of single cells.

6

Industrious Droplets for an Industrial Era

The Airbrush Paint Device, the Inkjet Printer, and Their Cousin the 3D Printer

Marshall Field's Wholesale Store was a landmark department store in Chicago that sold all kinds of "dry goods," such as textiles (from clothes and gloves to linens and upholstery), cutlery, clocks, and jewelry. Built in 1885 using an elegant stonework reminiscent of a Florentine *palazzo* (Figure 6.1), it stood seven stories and almost 40 meters tall with a two-level basement, totaling 12 acres of floor space—the equivalent of nearly seven soccer fields dedicated to

Figure 6.1 Marshall Field's Wholesale Store in 1885.

Image source: Chicagology.com. https://en.wikiarquitectura.com/wp-content/uploads/2019/05/Marshall-Filed-Wholesale-Store-2.png. Public domain image.

retail. Now gone, it remains a reference in American architecture and a testimony of the colossal commercial growth at the end of the nineteenth century. Ironically, microfluidics made the gargantuan job of painting its guts possible.

In 1887, Marshall Field's manager asked the building's maintenance supervisor, Joseph Binks, to paint its basement walls.[1] There were *miles* of them. The prospect of brush painting was daunting, so Binks decided to think out of the box. Without realizing it—although he quickly became aware of its potential—he devised a microfluidic solution.

This was the time when sprays were a big sensation. Dr. Solis-Cohen from Philadelphia was using Dr. Sales-Girons' *pulvérisateur*—based on the Rayleigh instability—and many spray models were already commercially available.[2] It would not be surprising if Binks might have heard about nebulized drugs and perfumes (see Chapter 3) or similar pneumatic inventions for spraying paint based on the Venturi effect.[3] In 1876, Francis E. Stanley, an artist from Kingfield, Maine, had created and patented an "atomizer" so he could spray "watercolors, India-ink or crayon and also for all kinds of shading in which color can be used in a liquid state."[4] Stanley's device never saw the market, but his Venturi-based design (Figure 6.2) is still used today. About 90 miles west of Chicago, in Rockford, Illinois, two brothers had started a company to manufacture the "airbrush," a modified version of Stanley's device.[a,5] It is just as likely that Binks may have also found inspiration from the way garden hoses create a spray when you pinch their end with your thumb.

Binks made a wand with a nozzle on the end and used a hand-operated pump to pressurize a vessel full of paint, ejecting the paint from the vessel through the wand (Figure 6.3, top). The presence of the nozzle—essentially a flat aperture approximately 1.5–2 millimeters in diameter—at the end of the wand

[a] Journals from that time reported that spray painting had been used in the railroad industry from the early 1880s.[6] In 1879, Abner Peeler, an inventor from Webster City, Iowa—credited with inventing an early typewriter and machines for engraving, knitting machine, and sewing—also invented the "paint distributer," a precursor of the airbrush that used a hand-operated compressor. In 1882, Peeler sold the patent to the Walkup brothers, who started the Airbrush Manufacturing Company in Rockford, Illinois in 1883. The name "airbrush" was coined by Liberty Walkup's wife Phoebe who was an accomplished watercolor artist. Other airbrush companies started around the same time as Binks' company, notably Thayer & Chandler (now Badger Airbrush Co.), Burdick (now Aerograph Company Ltd.), and Paasche (founded by an ex-employee of Thayer & Chandler, still in business).[3,5]

[1] Spray painting. *Wikipedia*. https://en.wikipedia.org/wiki/Spray_painting.

[2] Beatson, G. On spray producers. In *The Glasgow Medical Journal* (ed. Coats, J.), vol. 14. Glasgow and West of Scotland Medical Association, 1880.

[3] Airbrushdoc. Who invented airbrush? https://airbrushdoc.com/history/who-invented-airbrush (2018).

[4] Stanley, F. E. Atomizers. Google Patents. https://patents.google.com/patent/US182389A/en (1876).

[5] Merlin, B. The Airbrush Museum. https://www.airbrushmuseum.com (2021).

[6] Moser, W. The contentious historical origins of spray paint. *Chicago*. https://www.chicagomag.com/city-life/november-2011/the-contentious-historical-origins-spray-paint (2011, November).

fig: 1.

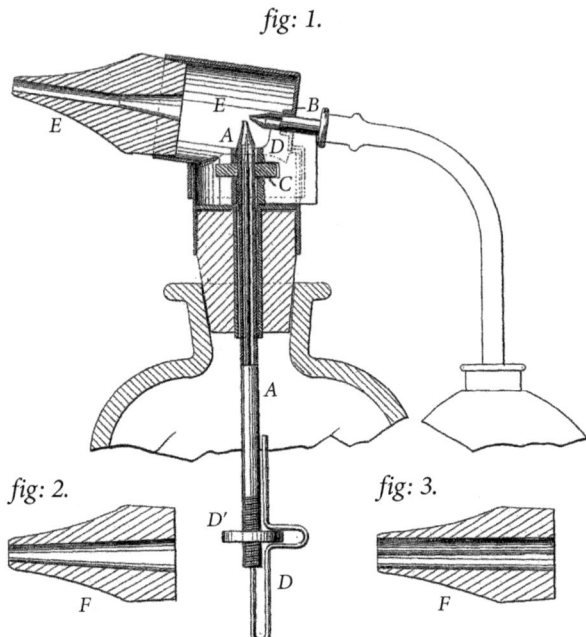

fig: 2.   *fig: 3.*

Figure 6.2 Francis Stanley's atomizer patent, a predecessor of the "airbrush."
Image source: Google Patents. https://patents.google.com/patent/US182389A/en. Public domain image.

caused the paint to eject as a high-velocity jet that quickly broke into a bouquet of droplets due to the Rayleigh instability. How clever of Binks.[b] The device made the task of painting large surfaces quick and, save for the hand pumping, effortless.

In a few years, Binks' invention—which he patented and commercialized under a company that still exists[c]—was catapulted to fame (Figure 6.3, top). In 1893, the city of Chicago became the site of a world's fair[d] that required repainting many of the public buildings. Within weeks of its inauguration, the fair's director was expecting millions of visitors, and more than 90% of the buildings remained to be painted when he learned of Binks' company and hired

---

[b] It is highly unlikely that Joe Binks, a building supervisor, would have known in 1887 about the explanation published by Lord Rayleigh in 1878—only nine years prior—in a specialized mathematical journal.[7]

[c] Bink's company, initially named The Star Brass Works, became Binks Sames and later Binks Manufacturing Co., which was eventually acquired by Carlisle Fluid Technologies, a global company that manufactures equipment for the application of sprayed materials such as paints. Carlisle also owns DeVilbiss.

[d] The World's Columbian Expo, held in 1893, celebrated the 400th anniversary of Columbus' arrival to the New World in 1492.

[7] Rayleigh, F. R. S. On the instability of jets. *Proc. London Math. Soc.* 10: 4–13 (1878).

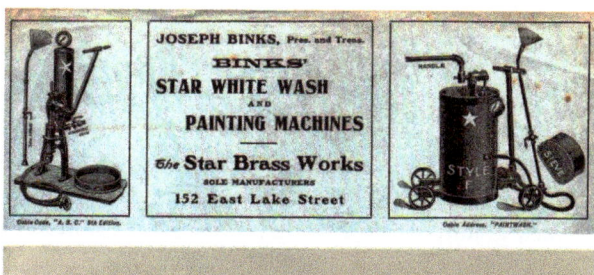

Figure 6.3 (Top) Advertisement for Joe Binks' spray painting company. (Bottom) Downtown Chicago in 1893, after being painted by Joe Binks' company.
Image sources: (Top) Chicagology.com. (Bottom) Wikimedia. Public domain images.

him. Binks' team of paint sprayers was able to spray whitewash over all the contracted buildings just in time for the opening. Because of this whitening, Chicago earned the nickname "The White City" (Figure 6.3, bottom). Before microfluidics even got its name, it did a lot for many people.

Industrialization brought with it the concentration of population in cities and the construction of new buildings—residential, offices, and factories—as well as all the furniture and vehicles. People had to paint all those surfaces. Spray painting was ideally suited for professional and industrial paint jobs, helping fuel economic growth. Almost every large surface you see evenly painted—without brush traces—has been sprayed with tiny paint droplets through a microfluidic nozzle.

One drawback of these airbrushes is that they require bulky, powerful pumps to displace the viscous paint. The portable version of Binks' spray machine was the *spray paint can*. (Confusingly, people also refer to it as an *aerosol spray* or *aerosol can*, which includes spray devices for many other liquids.) Half of the can contains paint, and the other half contains pressurized gas. A finger-actuated valve button directs paint toward the microfluidic nozzle. When the user presses the valve with the finger, the pressure buildup pushes the paint out through the nozzle in a high-speed jet, and the Rayleigh instability breaks the jet into a spray of paint droplets.

The impact of the aerosol cannot be understated. This tool is lightweight and inexpensive, and it became, as soon as it was introduced in 1949, a favorite of street artists to paint *graffiti* (Figure 6.4). Once considered a marginalized or outlaw form of art, graffiti art now sometimes achieves worldwide fame and top-earning figures, as is the case of the British (pseudonymous) artist Banksy. Manufacturers use the aerosol can format for many household chemicals, such as hairsprays, deodorants, air fresheners, and lubricants. More than 6 billion cans were manufactured worldwide in 2018, with the global aerosol can market reaching more than $10 billion.[8] Drop by drop, microfluidic nozzles have transformed our streets' appearance and modern life.

Airbrushes are used worldwide for art (Figure 6.5, left)—their original intended use[3]—and also to paint furniture, office spaces, millions of houses,[e] and every vehicle sold (Figure 6.5, right). Between 75 and 80 million vehicles worldwide are painted each year—more than 2 billion cars since 1900. All this

Figure 6.4 A graffiti artist shooting two cans of paint spray at once toward the wall.
Image source: Pxhere.com. https://pxhere.com/en/photo/1035366.

---

[e] In the U.S. alone, more than 1 million single-family homes are built yearly (source: U.S. Census Bureau).

[8] Mordor Intelligence. Aerosol cans market: Growth, trends, COVID-19 impact, and forecasts (2021–2026). https://www.mordorintelligence.com/industry-reports/global-aerosol-cans-market-industry (n.d.).

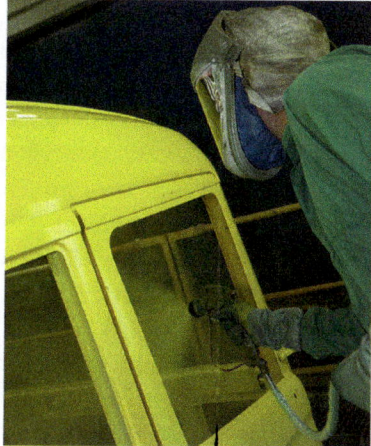

Figure 6.5 The airbrush (left), a tool that atomizes paint into airborne droplets. The tool is widely used by painters in the automotive industry (right).

Image sources: (Left) User Michel90 on Wikipedia. https://en.wikipedia.org/wiki/Airbrush#/media/ File:PaascheAirbrush.jpg (CC BY-SA 3.0). (Right) Pxhere.com. https://pxhere.com/en/photo/ 1391459.

bodywork demands that droplets between one-tenth and one-thousandth of a millimeter in diameter be uniformly deposited.[9] Sprays of minute droplets have fanned the massive efforts of industrialization.

* * *

If industrialization spurred the atomization of *paint* into droplets, the digital age would require the atomization of *ink* into droplets in the form of microfluidic inkjet printing. Both forms of atomization used—and still use—similar nozzles. You don't need to look any further. The book you are holding in your hands has been inkjet-printed so you can enjoy full-color images on every page in an affordable format.

It was not always like that. For more than 150 years, the publishing industry used offset printing, a technique based on patterned rollers that required very long runs to be cost-effective. But now that the costs and needs are shifting, presses are increasingly using inkjet printing to print books, or at least certain books, even though inkjet printing is more expensive on a per-unit basis. The publisher would have had to print tens of thousands of copies of this book for it to be cost-effective with traditional offset printing. If it weren't for microfluidics, the publishing industry would have collapsed.

---

[9] Poozesh, S., Akafuah, N., & Saito, K. Effects of automotive paint spray technology on the paint transfer efficiency: A review. *Proc. Inst. Mech. Eng. Part D J. Automob. Eng.* 232: 282–301 (2018).

Consider, for example, Scholastic—one of the top ten publishers globally and the largest publisher of children's books—which, up until 2019, printed all its books with traditional offset equipment in large runs. Offset printing requires a human operator to load the paper into the machine and program it. To meet the demand, Scholastic has shifted to runs of 3,000 books or less, which can be met by inkjet printing, even if at a higher unit cost.[10] By 2022, inkjet printing grew to 20% of the global print market. It has relegated offset printing to low-value, high-volume publications such as newspapers and magazines that are dwindling.[11] The rising trend of inkjet printing has been steady for decades. It's unstoppable.

Inkjet printing is more efficient because it is an automated microfluidic process. The printhead of an inkjet printer is a microfluidic device consisting of a nozzle that ejects tiny ink droplets. The droplets fly through the 1-millimeter air gap between the nozzle and the paper in a fraction of a thousandth of a second. Most inkjet printers spit out aqueous ink; some can print wax, giving a shiny photo appearance. Because the required setup time is minimal and a computer controls the whole process, the price of the total run can be much lower than with offset printing. Droplets can be automated, and automation is always cheaper.

It was the digital age that demanded the atomization of ink into droplets. It makes sense: In their electronic memories, computers internally express and manipulate data in "binary" bits of information ("0" or "1") as they turn tiny currents on or off. Breaking ink into small units makes it easier for computers to translate memory bits into ink bits, the binary language of the output.

Inkjet printing originates from a technique invented in the 1950s by a Swedish engineer named Rune Elmqvist, who wondered whether the phenomenon of droplet formation could be used in digital printing. Computers at the time had mechanical printer rolls and punch cards, which were annoyingly slow. Elmqvist worked for a company (later named Siemens) that was developing computers, so he wanted to find an automated way to rapidly print the outputs generated by the company's computers such as those of an electrocardiogram. Elmqvist knew about the Rayleigh instability: By forcing a liquid stream of aqueous ink at a relatively low speed through a nozzle that is just 50 micrometers (fifty-thousandths of a millimeter) wide, he could cause the jets of ink to break down swiftly into a single file of droplets flying through the air. He realized

[10] Ritarossi, L. Industry leaders talk book publishing & inkjet. *Inkjet Insight.* https://inkjetinsight. com/knowledge-base/industry-leaders-talk-book-publishing-inkjet (2022, January 25).
[11] Smyth, S. Digital printing to continue to take market share from offset presses. *Ink World.* https://www.inkworldmagazine.com/issues/2017-05-01/view_experts-opinion/digital-printing-to-continue-to-take-market-share-from-offset-presses (2017, April 13).

the droplets would acquire electrostatic charges in their friction with air. He devised a setup to deflect their trajectory on demand. In 1951, he built magnetic coils to generate a voltage he could control like an invisible wand directing the charged droplets to the right place.[12] The device, called a continuous inkjet printer (Figure 6.6) and considered the precursor of today's inkjet printers, included a reservoir and a pump for recycling the continuously flying droplets that had to be deflected away from the paper.

Continuous inkjet printers are still widely used in the manufacturing sector. They can quickly print from a considerable distance—up to 7 or 8 centimeters. The U.S. Food and Drug Administration mandates that manufacturers label consumer products with expiration dates, lot codes, and batch numbers. Mass-manufactured goods such as tomato cans, olive oil, and yogurts move around 5 meters per second on a conveyor belt. With continuous inkjet printers, manufacturers can print the mandatory codes on the packaging (Figure 6.7) in less than the blink of an eye. In 2019, the continuous inkjet printing market

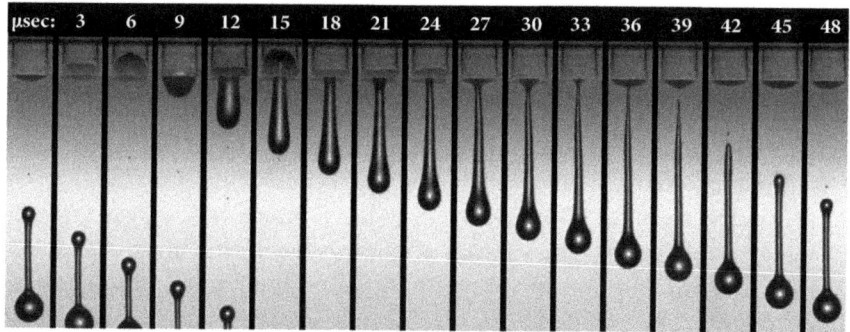

Figure 6.6 Sequence of micrographs depicting the process of ejection of an airborne ink droplet by the nozzle of a continuous inkjet printer. The diameter of the nozzle is 30 micrometers (30 thousandths of a millimeter), and the interval between each image is 3 microseconds (3 millionths of a second). The droplet is ejected at 9 microseconds. For these types of printheads, droplets can be as small as 1 picoliter (1 trillionth of a liter), they are produced at a rate between 10 and 100,000 droplets per second, and the final droplet velocity ranges from 5 to 10 meters per second (from 18 to 36 kilometers per hour—like a bicycle at full speed). Images adapted with permission from Arjan Fraters et al.[13]

---

[12] Elmqvist, R. Measuring instrument of the recording type. Google Patents. https://patents.google.com/patent/US2566443A/en (1951).
[13] Fraters, A., van den Berg, M., de Loore, Y., Reinten, H., Wijshoff, H., Lohse, D., et al. Inkjet nozzle failure by heterogeneous nucleation: Bubble entrainment, cavitation, and diffusive growth. *Phys. Rev. Applied* 12: 064019 (2019).

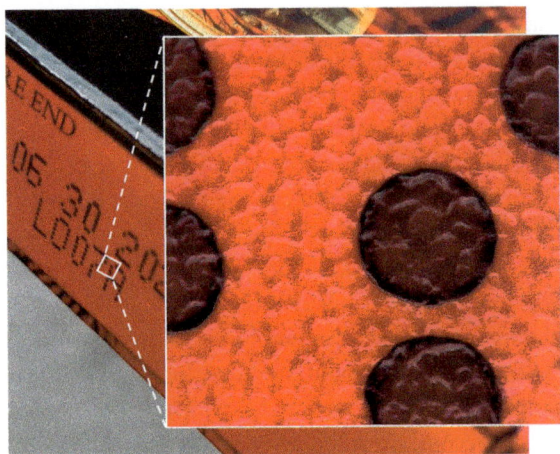

Figure 6.7 Expiration date printed on a cookie box with continuous inkjet printing. Each dot is approximately 100 micrometers (a tenth of a millimeter) wide. *Image by Albert Folch.*

accounted for more than one-third of the global inkjet printing market. Next time you check the expiration date on your milk carton, thank microfluidics for it.

The inkjet printers that you might have used in your office or at home to print a plane ticket or a family photo are the technological grandchildren of Rune Elmqvist's printer. Modern inkjet printers spit tiny ink droplets through a microarray of nozzles, like a bunch of Elmqvist's printers miniaturized and in a row. The nozzle arrays are microfabricated and shoot droplets on demand simultaneously, like a firing squad. In these modern printers, the nozzles only spit a droplet (a bit of ink) when instructed by the computer instead of a continuous train of droplets.

Microfluidic nozzle arrays were not developed from one day to the next. Making a tiny hole with a drill is easy, but upscaling the process to manufacturing thousands of exact holes and reliably controlling the ejection of each droplet is a very different story. Fortunately, born in the wake of the microelectronics revolution, the new field of MEMS (a convenient acronym for *microelectromechanical systems*) inspired the next generation of nozzles with integrated mini-controllers that could switch the nozzles on and off.

MEMS are similar to microelectronics and are fabricated with microelectronics technology, but they convert electrical energy into mechanical motion (as in modern computer disk drive heads) or vice versa within small-scale devices. To fabricate MEMS devices, researchers use a photoreduction technique inspired by photography dubbed *photolithography* to miniaturize microelectronics and MEMS devices (see Figure 2.4). In the dawn of the MEMS era, researchers had no choice but to use silicon or glass—the materials used in microelectronics—so they developed a set of techniques called *silicon micromachining* for precisely sculpting silicon. Using these methods, Jim Angell's lab at Stanford University had been building miniature sensors in silicon since the 1960s.[14,15] His work on MEMS and the new field of integrated circuits directly led to the first silicon-based microfluidic chip, utilizing gas as its fluid, and to the inkjet printer. MEMS are now used everywhere—to monitor blood pressure, measure acceleration in airbag deployers, read computer disk drives, project movies at movie theaters, and for many other applications—but it was a budding field back then.

It all started in 1964 when a neurologist visited Angell's lab and made an intriguing suggestion[16]: "Could this miniaturization technology be used to make things other than electronics, like smaller microelectrodes for brain record-ings?" Luckily for Angell, in 1965, an electrical engineering PhD student from Illinois named Kensall Wise joined his lab. Wise had worked for a year at Bell Labs and remembered they had been combining photolithography with chemical etching to carve channels in silicon. By 1970, Wise was routinely mak-ing neural probes in silicon as thin and narrow as the diameter of a small needle—the first MEMS device.[f,14]

Angell's lab was only a few miles northwest of NASA Ames Research Center. Because MEMS-based sensors were faster, smaller, and lighter than traditional sensors, NASA's program managers—who needed to reduce the weight of the onboard instruments and put life-monitoring, tetherless instruments on the astronauts—rushed to award research contracts to Angell.[17] Although Angell's devices never made it to space, his lab's MEMS technology would soon be applied to fabricate miniature printheads.

To achieve on-demand inkjet printing, the newly developed MEMS technol-ogy turned out to be ideally suited, as it allowed for the integration of electronics and tiny ink reservoirs next to the nozzles and thus control the ejection of every

---

[f] Wise eventually became a professor at Michigan, where he built a stellar research program based on his neural probes. Wise's probes are now used all over the world.

[14] Wise, K. D., Angell, J. B., & Starr, A. An integrated-circuit approach to extracellular microelec-trodes. *IEEE Trans. Biomed. Eng.* 17: 238–247 (1970).

[15] Samaun, Wise, K. D., Nielsen, E. D., & Angell, J. B. An IC piezoresistive pressure sensor for biomedical instrumentation. *IEEE Trans. Biomed. Eng.* 20: 101–109 (1973).

[16] Rafkin, L. The founder. *Forbes.* https://www.forbes.com/asap/2001/0402/MEMS_xtra1_print.html (2001, April 2).

[17] Butrica, A. J. NASA's role in the development of MEMS. In: *Historical Studies in the Societal Impact of Spaceflight* (ed. Dick, S. J), NASA SP-2015-4803, pp. 251–330. NASA, 2015. https://www.nasa.gov/sites/default/files/atoms/files/historical-studies-societal-impact-spaceflight-ebook_tagged.pdf.

drop. Ernest Bassous and Lawrence Kuhn, two engineers at IBM interested in printing technology, realized that the silicon geometries etched by Jim Angell's group at Stanford to make space-bound instruments would make perfect reservoirs for the inkjet nozzles they had in mind. A cavity or pit etched in silicon formed an inverted truncated pyramid with four smooth walls, always at the same angle (54.7°) due to the crystal structure of silicon (Figure 6.8). In the late 1970s, Bassous and Kuhn proceeded to fabricate a grid of pits, each with a bottom made of a thin, strong silicon nitride membrane. In the membrane, they produced holes by photolithography. This process created arrays of nozzles that a user could operate in parallel. With this first device, they could shoot 84,000 drops per second, each droplet flying a speed greater than 1 meter per second.[18] Importantly, the material used to make the pits—silicon—was very familiar to IBM researchers building microelectronics next door. Electronics could be integrated into the nozzle devices for the same price to control ink delivery. Integrating electronics with the nozzles made the printers faster, cheaper, and more precise.

Engineers have developed two methods to deliver ink on demand very fast and precisely (Figure 6.9, left). The *piezoelectric nozzle* uses a flat piece of material that bends when electrically excited, a bit like the muscle of a squid or a scallop when it jets out water to propel itself. A sudden, brief burst in voltage

Figure 6.8 The crystal structure of silicon is manifested when it is etched with certain etchants.

Image courtesy of CBMS (Okaar Photography). Device made by Andreas Manz ca. 1989.

[18] Bassous, E., Taub, H. H., & Kuhn, L. Ink jet printing nozzle arrays etched in silicon. *Appl. Phys. Lett.* 31: 135–137 (1977).

PIEZOELECTRIC   BUBBLEJET

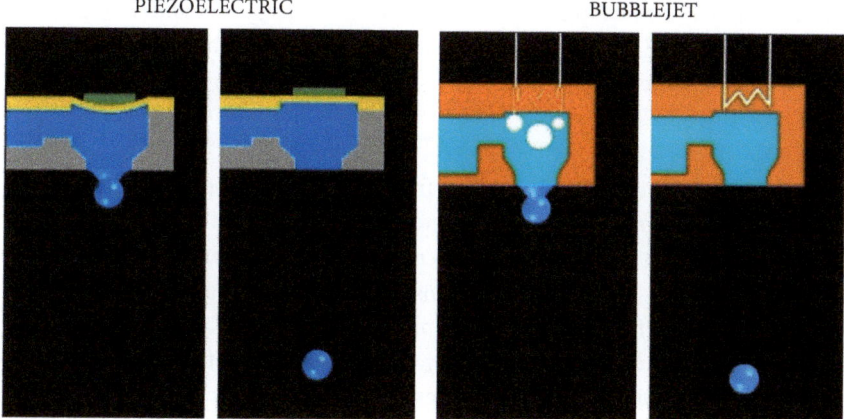

Figure 6.9 Schematic representation of the two main methods for droplet generation in inkjet printers. (Left) The direct displacement of fluid by a piezoelectric actuator causes the ejection of a droplet. (Right) In the BubbleJet method, applied heat vaporizes the ink and forms one or more bubbles, which causes the ejection of a droplet.

Image source: Screenshots from movie by user Javachan on Wikipedia. https://en.wikipedia.org/ wiki/Inkjet_printing#/media/File:Micro_Piezo_Comparison.gif. Fair use.

applied to the piezoelectric material causes the ejection of a tiny amount of fluid through the nozzle. The nozzle can repeat this process thousands of times a second.

The other nozzle type is a lightning-speed, miniature ink boiler conceived by Japanese engineer Ichiro Endo at Canon in 1977. Endo realized he could generate a bubble by heating the ink in less than a millisecond to its boiling point, thus displacing enough ink in a reservoir to eject a single droplet (Figure 6.9, right).[19] His prototype became the "BubbleJet" Canon printer. The nozzle can work ultra-fast simply because it is ultra-small, less than a billionth of a liter. By comparison, it typically takes minutes to bring the familiar pasta pot to a boil. Not only does it contains a lot of water but also water is a substance that does not absorb heat easily. In the BubbleJet printer, the volume of ink that needs boiling is about a billion times smaller than a pot of water. In addition, the BubbleJet boiler pot—the reservoir with slanted walls—is made of silicon, a material that conducts heat almost 400 times faster than water, behaving like an ice pack that quickly absorbs the ink's heat and enables lightning-fast control of the printing heat cycle. Endo had figured it all out. As incredible as it may sound, a BubbleJet

---

[19] Nielsen, N. J. History of ThinkJet printhead development. *Hewlett-Packard J.* 36(5): 4–10 (1985). http://www.hpl.hp.com/hpjournal/pdfs/IssuePDFs/1985-05.pdf.

printer can turn the heat on and off thousands of times per second. For students and experts alike, Endo's accomplishment remains one of the most astonishing feats of microfluidic engineering.

* * *

Microfluidic advances have also come from *preventing* droplet breakup as fluids go through a nozzle. The only way to do that is by modifying the ominous surface tension. In 1988, mechanical engineer Scott Crump wondered if the ink was so viscous that the surface tension was not strong enough to slice it into droplets, and *filament* shapes would result instead. Could the nozzle be programmed to move and deposit layer after layer of different materials to build three-dimensional (3D) structures—a *3D printer*? Crump heated plastic and extruded it into a filament through a nozzle, much like a pasta machine extrudes spaghetti or capellini. This technique is known as *fused filament fabrication* (FFF) or *fused deposition modeling* (FDM) (Figure 6.10). If you have seen a 3D printer in your children's school or own one at home, it's likely the type invented by Crump. The molten plastic capellini hardens by spontaneous cooling immediately after extrusion.

The filament, which can be made of various materials, is usually fed from a large spool into the nozzle. A digital design file instructs the printer to build the print layer by layer, moving the printhead in X, Y, and Z directions as it "reads" the file. The nozzle moves in all three X, Y, and Z directions under computer control with the help of gears, ribbons, and electric motors. The nozzle's diameter determines the resolution, typically between 0.3 and 1 millimeter, and cannot be altered once the print starts. There are many other types of 3D printers, but these are undoubtedly the cheapest and the most user-friendly.

3D printers can replace traditional manufacturing technologies, so we become less dependent on them. Imagine you are stranded in a remote location—such as an island in the middle of the ocean or, in the future, on the moon or Mars—and need to repair a mechanical part or an instrument. Instead of having the part shipped to you, which could take months, having a 3D printer in place would be more efficient. You could quickly 3D print a replacement on-site simply by downloading the file. The International Space Station, collaborating with pioneer startup Made In Space, installed an FDM 3D printer in 2014 to replace mechanical parts in space. On December 11, 2014, NASA astronaut Yvonne Cagle used the 3D printer to print a buckle, the first functional object ever manufactured off the surface of our planet. The good news is that the mechanical properties of the parts are not affected by microgravity.[20]

---

[20] Gaskill, M. Solving the challenges of long duration space flight with 3D printing. NASA. https://www.nasa.gov/missions/station/solving-the-challenges-of-long-duration-space-flight-with-3d-printing (2019).

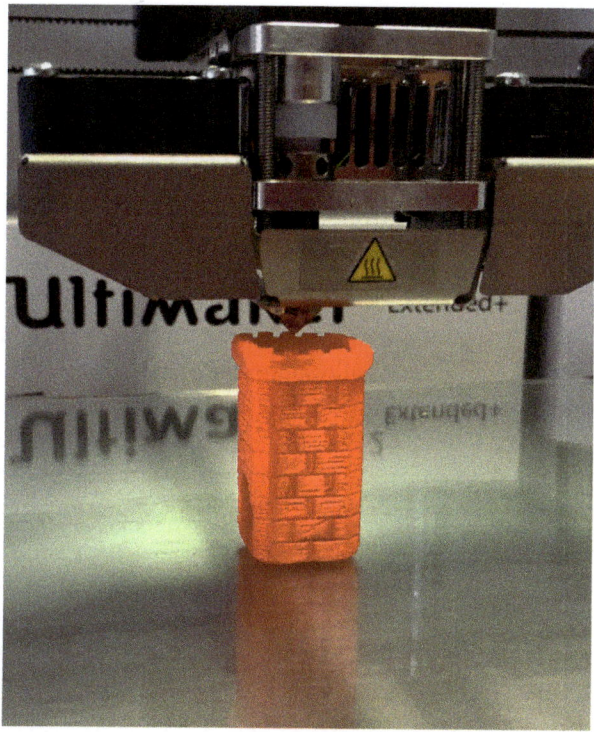

Figure 6.10 Close-up of the printhead of a fused filament fabrication 3D printer. The filament is red. The nozzle is the inverted metallic cone at the top of the printed medieval tower. The notched ribbons that guide the motion of the printhead are visible at the top of the image.
Image source: Pxhere.com.

Down on Earth, the widespread availability of these 3D printers enables an approach to manufacturing called *distributed manufacturing* or *cloud manufacturing*,[21] a means of production in which the fabrication of products is distributed over a network of small-scale, decentralized nodes. Cloud manufacturing uses digital designs that multiple teams working in parallel can share. Hence, the approach is very efficient for the iteration of prototypes and even more so for the remarkably easy-to-use FDM 3D printers. Unlike mass manufacturing based on rigid supply chains, cloud manufacturing relies on agile small manufacturers responding swiftly to shifting inventories and market demands.[22] During the COVID-19 pandemic, cloud manufacturing of health care equipment partially compensated for the manufacturing and distribution shortages, as Chiari villagers recall quite well.

---

[21] Xu, X. From cloud computing to cloud manufacturing. *Robot. Comput. Integr. Manuf.* 28: 75–86 (2012).
[22] Lipson, H., & Kurman, M. *Fabricated: The New World of 3D Printing.* Wiley, 2013.

Chiari is a tiny town of red-tile roof houses encircling its cathedral at the center of Lombardy in northern Italy. In March 2020, Italy became the first Western country to declare a nationwide lockdown in an attempt to contain the outbreak of COVID-19. This tiny virus travels from host to host inside invisible droplets of saliva or nasal mucus. At the end of March, the small hospital outside Chiari was about to reach its saturation point for new patients in the emergency room. In Bergamo, just half an hour by car northwest of Chiari, its only crematorium was working 24/7, and a parade of military trucks had to transport coffins to nearby cities.[23]

That year, the virus plunged the world into a global pandemic and economic standstill. People were not allowed to leave their homes, let alone travel. Offices, restaurants, schools, theaters, and many businesses closed their doors or switched to the online format. Our social life shrank onto a flat screen as people disengaged by muting themselves and turning off their webcams. The manufacture and distribution of materials and devices—the infamous *supply chain*—for various essential services, particularly health care, came to a grinding halt. It was as if the world had shut down.

One day, the Chiari hospital staff realized they were running out of respirator valves for the COVID-19 patients in the intensive care unit. Someone decided to issue a desperate call for help through the *Giornale di Brescia*, a local newspaper. Brescia is just 30 km east of Chiari. The distress message was read by physicist Massimo Temporelli, founder of a Milan-based 3D printing startup.[24] Milan is 70 km west of Chiari, on the same road but opposite Brescia.

Confined at home in Milan, Temporelli Googled sister startups near Chiari and, in a few minutes, found one in Brescia called Isinova. Temporelli contacted the company, and within 6 hours, Isinova's Cristian Fracassi and Alessandro Ramaioli collaborated with Temporelli to reverse-engineer the original $11,000 device. They began manufacturing the valves—for approximately $1 each (Figure 6.11)—which were delivered and used on ten patients by the end of the day.[24] In a matter of hours, news outlets ran the story worldwide, and the Italian Minister of Health congratulated the scientists for their speedy accomplishment.[25]

[23] Fasano, S. Coronavirus: Frontline hospital at saturation point in Italy: Doctor. *Alarabiya News*. https://english.alarabiya.net/features/2020/03/20/Coronavirus-Frontline-hospital-at-saturation-point-in-Italy-doctor (2020, March 20).

[24] Zarzalejos, A., & Moynihan, Q. A startup 3D-printed emergency breathing valves for COVID-19 patients at an Italian hospital in less than 6 hours. *Business Insider*. https://www.businessinsider.com/coronavirus-italian-hospital-3d-printed-breathing-valves-COVID-19-patients-2020-3 (2020).

[25] Molitch-Hou, M. 3D printing for COVID-19, Part Two: Spare valves for oxygen masks. 3dprint.com. https://3dprint.com/265022/3d-printing-for-covid-19-part-two-spare-valves-for-oxygen-masks (2020, March 27).

Figure 6.11 3D printed respirator valves.

Image courtesy of Massimo Temporelli, Università degli Studi di Milano and President of The FabLab.

Similar stories popped up around the world. Dr. Albert Chi, a trauma surgeon from Oregon Health and Science University, needed ventilators for his patients, and the supplier could not deliver during the pandemic. He contacted Limbitless Solutions, a 3D printing prosthetics company in Florida, which designed a 3D printable ventilator. Not only do its 3D printed parts cost less than $10 apiece and take just 3 hours to print but also the device does not require electricity or electronics and works only with the airflow from an oxygen tank, so it can be manufactured in remote places and used almost anywhere in the world, even in areas without electricity.[26]

The pandemic made people realize that computer-assisted forms of manufacturing such as 3D printing—collectively called *digital manufacturing*—are faster, cheaper, and often better. Medical and testing devices from valves to swabs, personal protective equipment from face masks to adjusters and fitters, training and visualization aids, door openers, you name it—thousands of users spontaneously shared their designs online. They contributed their 3D printers

---

[26] Molitch-Hou, M. 3D printing and COVID-19, May 12, 2020 update: Limbitless solutions, Dunlee, 3DPRINTUK. 3dprint.com. https://3dprint.com/267257/3d-printing-and-covid-19-may-12-2020-update-limbitless-solutions-dunlee-3dprintuk (2020).

to make these parts for hospitals.[27,28] Although the press celebrated these stories widely and generously, an important detail was left out: The critical component of these printers was—and is—microfluidic nozzle.

A sister technique of FFF is *direct ink writing*. The only difference is that the capellini or ink filament is not created by heat; instead, pressure applied inside the nozzle extrudes a slurry or a paste, generally at room temperature, much like you squeeze toothpaste out of the tube. A system of valves can switch from one slurry or paste to the next using different reservoirs operated in parallel. Some pastry chefs use food 3D printers to build custom-made cakes with fancy shapes because (warm) chocolate and other foods make good slurries (Figure 6.12). Some artists use 3D printers to print ceramics. Electrical engineers have miniaturized this concept and used metallic viscous solutions to print LEDs,[29] batteries,[30] antennas,[31] and electrodes within biological tissue.[32] These flexible metallic patterns can be curled around your wrist to produce *wearables*.

Engineers have fashioned a type of direct ink writer called a *bioprinter* that extrudes biological inks, or *bioinks*, often containing live cells. These bioinks are like the gelatin you may eat as a dessert or the gel people use for shaving—a type of half-solid–half-liquid material that scientists call *hydrogel*. Hydrogels usually have a very high viscosity, so they are fairly immune to droplet breakup; some are elastic or have lower viscosity when they flow. Some of the best bioprinters in the world are in Jennifer Lewis' lab at Harvard University, where researchers have developed special nozzles and hydrogel bioinks to quickly switch ink and produce multicolored ink "capellini" (Figure 6.13, top).[33]

Lewis also devised an alternative way of modifying the surface tension to prevent droplet breakup. One day, she was in the lab with her student, James Smay, trying a slurry type of ink, and she saw how he was moving the printhead through a reservoir to prevent the ink from drying. "I noticed that the residual pressure within the nozzle led to the extrusion of an ink filament that maintained

---

[27] Choong, Y. Y. C., Tan, H. W., Patel, D. C., Choong, W. T. N., Chen, C. H., Low, H. Y., et al. The global rise of 3D printing during the COVID-19 pandemic. *Nat. Rev. Mater.* 5: 637–639 (2020).

[28] Radfar, P., Razavi Bazaz, S., Mirakhorli, F., & Warkiani, M. E. The role of 3D printing in the fight against COVID-19 outbreak. *J. 3D Print. Med.* 5: 51–60 (2021).

[29] Kong, Y. L., Tamargo, I. A., Kim, H., Johnson, B. N., Gupta, M. K., Koh, T. W., et al. 3D printed quantum dot light-emitting diodes. *Nano Lett.* 14: 7017–7023 (2014).

[30] Sun, K., Wei, T. S., Ahn, B. Y., Seo, J. Y., Dillon, S. J., &Lewis, J. A. 3D printing of interdigitated Li-ion microbattery architectures. *Adv. Mater.* 25: 4539–4543 (2013).

[31] Adams, J. J., Duoss, E. B., Malkowski, T. F., Motala, M. J., Ahn, B. Y., Nuzzo, R. G., et al. Conformal printing of electrically small antennas on three-dimensional surfaces. *Adv. Mater.* 23: 1335–1340 (2011).

[32] Mannoor, M. S., Jiang, Z., James, T., Kong, Y. L., Malatesta, K. A., Soboyejo, W. O., et al. 3D printed bionic ears. *Nano Lett.* 13: 2634–2639 (2013).

[33] Hardin, J. O., Ober, T. J., Valentine, A. D., & Lewis, J. A. Microfluidic printheads for multimaterial 3D printing of viscoelastic inks. *Adv. Mater.* 27: 3279–3284 (2015).

Figure 6.12 Examples of direct ink writing (DIW). (Top) A DIW printer in the process of 3D printing chocolate. (Bottom) A 3D printed chocolate rose.
Images courtesy of Cocoa Press.

its shape within this liquid phase," she recalled.[34] It was a Eureka moment. They developed a 3D printer where the inks are extruded not into air but instead into a gel-like viscous matrix, effectively canceling the surface tension.

[34] Folch, A. *Hidden in Plain Sight: The History, Science, and Engineering of Microfluidic Technology.* MIT Press, 2022, p. 266 and 269–270.

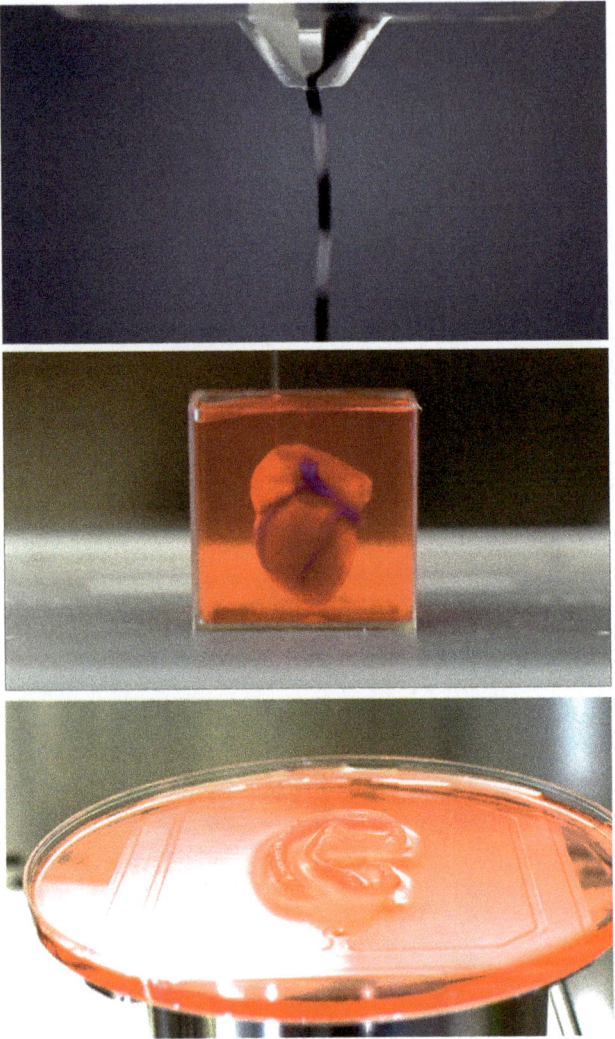

Figure 6.13 Examples of multimaterial 3D printing and bioprinting.

(Top) Multimaterial 3D printing with a direct ink writer. The extrusion of a multimaterial thread is achieved by quickly switching the ink with valves inside the nozzle. (Middle) A miniature heart with its own vasculature in the process of being 3D printed with human cells using an omnidirectional bioprinter. The nozzle of the printer is visible as a thin needle half submerged in a gelatinous red fluid. (Bottom) A 3D printed ear.

(Top) Image courtesy of Jennifer Lewis, Harvard University. (Middle) Image courtesy of Tal Dvir, Tel Aviv University. (Bottom) Image courtesy of Dan Cohen, 3DBio Therapeutics.

With her *omnidirectional printer*, the nozzle can 3D print in all directions and can print soft structures and materials that would otherwise collapse.[31,35] Using

---

[35] Ahn, B. Y., Duoss, E. B., Motala, M. J., Guo, X., Park, S.-I., Xiong, Y., et al. Omnidirectional printing of flexible, stretchable, and spanning silver microelectrodes. *Science* 323: 1590–1593 (2009).

Lewis' omnidirectional printing approach, a team led by Tal Dvir printed a miniature heart and its vasculature by feeding the printer two bioinks made of genetically reprogrammed cells. At the end of the print, the mini-heart was suspended in a gel (Figure 6.13, middle).[36] One ink contained heart muscle cells and the other blood vessel–forming cells. Although the heart is the size of a rabbit's and cannot pump fluids yet, the muscle cells did develop electrical contractility. "This is the first time anyone anywhere has successfully engineered and printed an entire heart replete with cells, blood vessels, ventricles, and chambers," said Tal Dvir.[34] These 3D-printed bioinks and cells are not expected to cause an immune reaction because they are derived from the patient's cells. With these printers, called *bioprinters*, scientists hope one day to be able to print live organs (or part of them) that can replace damaged ones.[37]

There have been some spectacular successes already. In 1999, surgeon–bioengineer Anthony Atala's group at Boston Children's Hospital (now at Wake Forest University) successfully used the bladder cells of 10-year-old patient Luke Massella to bioprint a bladder replacement. Luke has grown with the bioprinted bladder since then. "It was pretty much like getting a bladder transplant, but from my own cells, so you don't have to deal with rejection," he said. "I was able to live a normal life after."[38] Although the microfluidic basis of bioprinters is rarely mentioned, Atala's breakthrough bioprinting work has been recognized by publications such as *TIME* magazine, *Scientific American,* and *U.S. News & World Report.* Microfluidics is silently making a difference.

Not just academic teams like Atala's but also biotech companies are starting to reach the clinic. On June 2, 2022, a New York company called 3DBio Therapeutics summoned journalists to its Queens headquarters for a press conference. It was a nice Sunday and nothing in the surrounding streets foretold the importance of the moment. Neighborhood life was bustling as usual with the taxi honks, the colors of the grocery stores, and the pizza smells. But biotech companies do this when they have something important to announce, so there was a lot of expectation for the event. Amid camera flashes and microphones, Dr. Cohen, the chief executive officer, proclaimed the successful implantation to a 20-year-old female patient of a bioprinted ear.[39]

[36] Noor, N., Shapira, A., Edri, R., Gal, I., Wertheim, L., & Dvir, T. 3D printing of personalized thick and perfusable cardiac patches and hearts. *Adv. Sci.* 6: 1900344 (2019).

[37] Wu, Y., Yang, X., Gupta, D., Alioglu, M. A., Qin, M., Ozbolat, V., et al. Dissecting the interplay mechanism among process parameters toward the biofabrication of high-quality shapes in embedded bioprinting. *Adv. Funct. Mater.* 34: 2313088 (2024).

[38] Lord, B. Bladder grown from 3D bioprinted tissue continues to function after 14 years. 3D Printing Industry. https://3dprintingindustry.com/news/bladder-grown-from-3d-bioprinted-tissue-continues-to-function-after-14-years-139631 (2018, September 12).

[39] Lewis, S. 3DBIO Therapeutics and the Microtia Congenital Ear Deformity Institute conduct human ear reconstruction using 3D bioprinted living tissue implant in a first-in-human clinical trial. Business Wire. https://www.businesswire.com/news/home/20220602005051/en/3DBio-Therapeutics-and-the-Microtia-Congenital-Ear-Deformity-Institute-Conduct-Human-Ear-Reconstruction-Using-3D-Bioprinted-Living-Tissue-Implant-in-a-First-in-Human-Clinical-Trial (2022, June 2).

Cohen explained that the ear had been bioprinted using cells taken from the patient herself (Figure 6.13, bottom). The woman was born with a small and misshapen left ear—a condition called microtia. People born with microtia usually have an intact inner ear, but they experience hearing loss due to the missing external structure. Plastic surgeons sometimes build replacement ears from a patient's harvested rib or with materials that have a Styrofoam-like consistency. These options can be challenging or painful, and the ears rarely look completely natural or perform well.

Cohen's journey with printing human cells dates back almost twenty years when he was an undergraduate in the Lipson lab, then at Cornell University. Plastic surgeon Jason Spector and biomedical engineer Lawrence Bonassar realized that Hod Lipson's 3D printers could be used to quickly replicate cell-seeded hydrogels into clinically realistic shapes.[40] The researchers decided to specifically work on the 3D printing and replacement of cartilage because cartilage structures—such as joints, the spine, and the nose—are not vascularized with blood vessels. Absent the intricate vasculature, cartilage tissue appeared simpler to 3D print than other tissues. The challenge was reduced to taking 3D pictures of a human ear with a scanner, mixing cartilage cells with a biological ink called hydrogel, and feeding the instructions to the printer so it could squirt the ear's shape layer by layer from the 3D image files. In 2013, they published the first experiments with rat cells. As the cells grew in the hydrogel for a few weeks, they assembled into cartilage tissue with the desired ear shape.[41]

Not all 3D printers work by extrusion through a nozzle. Some 3D printers have inkjet nozzles that can deposit low-viscosity polymers droplet by droplet in air. These special polymers—called *photocurable resins*—solidify rapidly when ultraviolet light hits them.[42] Unlike other 3D printers, they can print multiple polymers simultaneously or in sequence (Figure 6.14). The printer activates different nozzles in parallel, feeding each nozzle from a different ink reservoir. However, although these inkjet-based 3D printers can print directly in air, they are still limited and cannot yet re-create materials, for example, with the properties of glass, paper, or biological tissue.

The nozzle appears as a common theme for generating colored droplets in many industrial applications, from the spray painting apparatus devised by Joe

---

[40] Nutt, D. Cornellian-founded company implants 3D-bioprinted ear. *Cornell Chronicle.* https://news.cornell.edu/stories/2022/06/cornellian-founded-company-implants-3d-bioprinted-ear (2022, June 2).

[41] Reiffel, A. J., Kafka, C., Hernandez, K. A., Popa, S., Perez, J. L., Zhou, S., et al. High-fidelity tissue engineering of patient-specific auricles for reconstruction of pediatric microtia and other auricular deformities. *PLoS One* 8: e56506 (2013).

[42] Naderi, A., Bhattacharjee, N., & Folch, A. Digital manufacturing for microfluidics. *Annu. Rev. Biomed. Eng.* 21: 325–364 (2019).

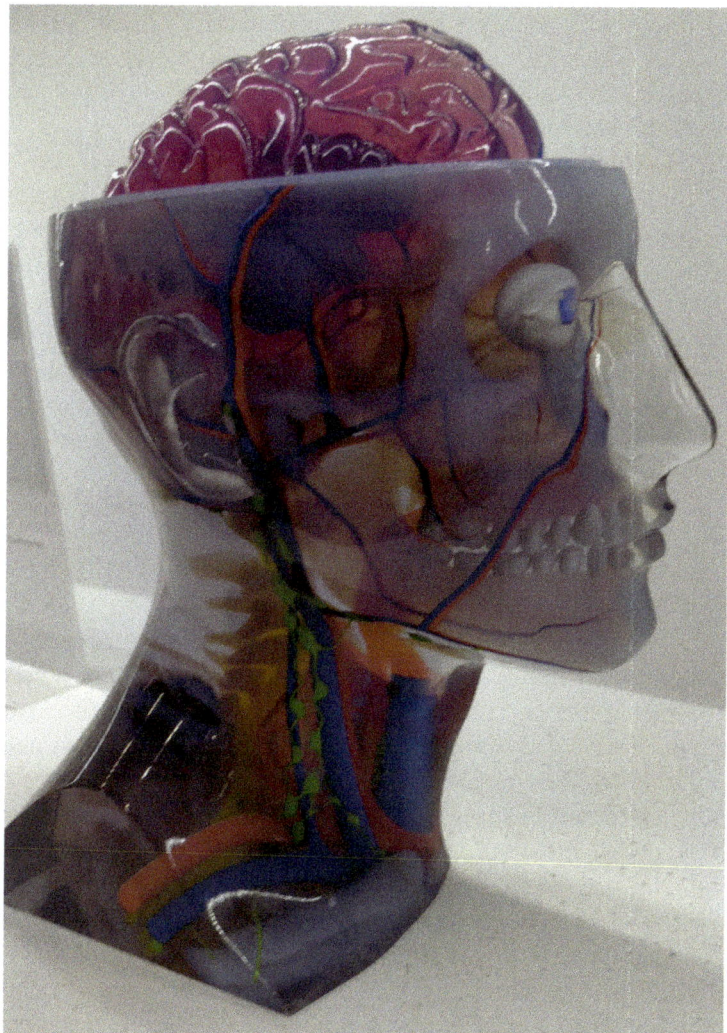

Figure 6.14 This 3D model of a skull and head for an anatomy class was 3D printed with a multimaterial inkjet printer from a single digital file that specifies all the components.

Image source: User Florian.Kohn on Wikipedia. https://en.wikipedia.org/wiki/Multi-material_ 3D_printing#/media/File:Material_Jetted_Model_of_a_Human_Skull.jpg (CC BY-SA 4.0).

Binks in 1887 to the continuous inkjet printer invented by Rune Elmqvist in 1951—and it functions better as it gets smaller. Like most MEMS and microelectronics devices, the microfluidic nozzle benefits from cost reduction by batch fabrication and integration. A modern inkjet printer fits thousands of nozzles in a thumbprint-sized area, organized in linear arrays (usually one or two lines for each ink color), with nozzles as small as 9 micrometers.[43] The printer shoots individual ink drops that are in the range of approximately 3.5–5 picoliters (1 picoliter is one-millionth of a millionth, or a trillionth, of a liter, a volume similar to that of a bacterium). The printed spot sizes on paper are in the range of 10–20 micrometers in diameter—that's five or ten times thinner than one of your hairs. As the nozzles are scaled down, their ink consumption and speed performance are improved, and more can be packed in one print head. They also become less expensive to fabricate per unit, so the market keeps growing.[g]

The old miniaturization law has prevailed for the nozzle: The smaller, the faster, the cheaper, the better. This tiny microfluidic component, born in the dawn of microelectronics, is as inconspicuous as it is essential for the proper functioning of millions of printers and 3D printers. Microfluidic nozzles have allowed for printing every image and word on these pages down to this dot, here, this one.[h]

Summary

- Like all spray devices, paint spray devices can be classified according to whether they are based on the *Venturi effect* (a constriction in an air pipe generates a suction effect that ejects paint from a reservoir, and the collision between the paint stream and the air current generates paint droplets) or on the *Rayleigh instability* (the liquid's surface tension slices the paint jet into droplets after it is ejected through a nozzle).
- The *airbrush*, based on the Venturi effect, was invented by Francis Stanley in 1876. It was commercialized by the Walkup brothers, who founded the Airbrush Manufacturing Company in Rockford, Illinois, in 1883. The name "airbrush" was coined by Liberty Walkup's wife Phoebe, who was an accomplished watercolor artist.
- In 1887, Joe Binks, a building maintenance supervisor in Chicago, came up with a way to pressurize paint with a pump through a small nozzle, which causes the breakdown of the jet of paint into droplets due to the Rayleigh instability.

---

[g] In 2019, the global inkjet printer market was valued at more than $34 billion with a 5% growth rate.

[h] The last words of this chapter are a tribute to Italian writer and chemist Primo Levi. During his imprisonment at Auschwitz, he dreamt of a beautiful story about the fate of a single carbon atom that became *Carbon*—famously ending in "[...] *this dot, here, this one.*"—, the last chapter of his 1975 best-seller *The Periodic Table*. The Royal Institution of Great Britain named it The Best Science Book Ever Written.

[43] Canon. Technology used in inkjet printers: FINE (full-photolithography inkjet nozzle engineering). https://global.canon/en/technology/support06.html (2018, December 27).

- Paint droplets have been essential for the industrialization of society. Paint-spraying devices were and continue to be used for painting the large surfaces of buildings, vehicles, and furniture, among others.
- *Aerosol spray cans* are the portable version of the device invented by Joe Binks and have had a major impact in art and decoration.
- *Continuous inkjet printing* and *inkjet printing* are based on the deposition of droplets through a microfabricated nozzle.
- Fused deposition modeling, also known as fused filament fabrication (FFF), is a form of 3D printing based on the extrusion of a molten filament through a microfluidic nozzle.
- During the COVID-19 lockdown, 3D printing allowed for printing essential medical parts and thus bypassing shortages in the supply chain.
- Bioprinting is a form of 3D printing similar to FFF that can be used to print and replace live tissues.

PART II

WICK IT UP

If you have the knowledge, let others light their candles in it.

—Margaret Fuller (1810–1850)

Wicking, the spontaneous penetration of fluids in small spaces, tubes, or pores, is a fundamental process in our world. Because scientists historically studied the phenomenon in *capillaries* or tubes as thin as a hair (*capillus* in Latin), it is also known as *capillarity* or *capillary action*. Capillarity is a pump that does not require batteries or other external energy sources. Without it, plants would not grow; soil would not be fertile; candles and oil lamps would not light up; and gauzes, pads, tissues, rags, and napkins would not absorb unwanted fluids for us. Capillarity is omnipresent.

The list of scientists who have paid attention to capillarity is long and illustrious. Pliny the Elder, the Roman naturalist who died in Pompeii during the eruption of Mount Vesuvius in 79 AD, wrote about soaking papyrus with extracts by capillary action to detect impurities.[1] Leonardo da Vinci did the first formal experiments in capillarity toward the end of the fifteenth century. In 1605, the English philosopher Francis Bacon, considered one of the developers of the scientific method, noted in his influential book, *Thoughts on the Nature of Things*, that when a sugar cube was dipped into wine, the wine would rise up the cube. He was not inebriated—he was pondering the atomistic nature of matter, a controversial subject at the time. In 1664, the Anglo-Irish physicist Robert Boyle, who is also widely recognized as the first modern chemist, made his discovery that violet syrup could act as a "chemical indicator"—that is, it would turn green or red depending on whether it was dipped in a base or an acid, respectively, by soaking the syrup in a strip of paper.[2]

But capillarity presented a series of confounding riddles: Why did liquids move spontaneously, even against gravity? Why did they prefer to rise higher

---

[1] Yagoda, H. Applications of confined spot tests in analytical chemistry: Preliminary paper. *Ind. Eng. Chem. Anal. Ed.* 9: 79–82 (1937).
[2] Szabadváry, F. Indicators: A historical perspective. *J. Chem. Educ.* 41: 285–287 (1964).

in thinner tubes? The explanation eluded minds such as British scholar Robert Hooke, a contemporary of Boyle, who erroneously thought it was due to the greater difficulty of atmospheric gases penetrating narrower tubes, thus exerting less pressure. In 1666, Dutch librarian and polymath Isaac Vossius unveiled the currently accepted explanation. Capillarity, Vossius said, was a phenomenon triggered by the water-loving property of the glass surface (an affinity now called *hydrophilicity*) drawing the water up the tube. Thus, narrower tubes presented proportionally more points of contact of adherence to the water with respect to the ascended volume, leading to a higher rise.[3] In 1706, an assistant of Sir Isaac Newton named Francis Hauksbee dipped glass tubes of different diameters into water. He confirmed experimentally for the first time that liquids rose higher in the smaller diameter glass tubes.[4] A few years later, James Jurin noted that the height a fluid rises in a capillary column was determined only by the cross-sectional area at the surface, not by any other column dimensions.[5] Students now learn this empirical finding as *Jurin's law*. Once the theory was developed further, at the beginning of the nineteenth century, Thomas Young and Pierre Simon-Laplace derived the equation for capillary action named after them.[6,7] And in 1901, a 22-year-old PhD student at the University of Zurich named Albert Einstein—who would shake the world with his *theory of relativity* four years later—published his first paper, "Consequences of the Observations of Capillarity Phenomena."[8]

How can liquids rise spontaneously and all the higher for narrower tubes? As surmised by Vossius, the "engine" of capillarity is the surface's water-loving property: Water likes to wet the surfaces of glass tubes. Hence, the fluid crawls up the walls (both inside and outside the tube), forming a *meniscus*. But as soon as the meniscus has formed, something happens inside the capillary that does not occur outside. The elastic surface of the water will stretch, and the water surface tension will tend to relax the surface to a more stable position, drawing the column of water up (Figure PII.1). Next, the edges will wet the walls and crawl up again, successively. The motion only stops when the "push force" exerted by the elastic surface equals the "pull weight" of the water column. In narrow tubes, the

---

[3] Vossius, I. *De Nili et aliorum fluminum origine [On the Origin of the Nile and Other Rivers]*. Den Haag (Adrian Vlacq), 1666.

[4] Hauksbee, F. An experiment made at Gresham-College, showing that the seemingly spontaneous ascension of water in small tubes open at both ends is the same in vacuo as in the open air. *Philos. Trans.* 25: 2223–2224 (1706).

[5] Jurin, J. An account of some experiments shown before the Royal Society; With an enquiry into the cause of the ascent and suspension of water in capillary tubes. *Philos. Trans.* 30: 739–747 (1718).

[6] Young, T. An essay on the cohesion of fluids. *Philos. Trans. R. Soc. London* 95: 65–87 (1805).

[7] Laplace, P. S. Supplément au dixième livre du Traité de Mécanique Céleste. *Traité de Mécanique Céleste* 4: 1–79 (1805).

[8] Einstein, A. Folgerungen aus den Capillaritätserscheinungen. *Ann. Phys.* 309: 513–523 (1901).

Figure PII.1 Spontaneous wicking of water into a glass capillary.
Image by Albert Folch.

weight of the water column is much smaller than that in wider tubes, and the effect of the force exerted by the elastic surface is proportionately stronger—that's why liquids rise higher in narrower capillaries. Wicking, therefore, is the process by which the energy stored *within the liquid* propels the liquid through narrow spaces. Strictly speaking, it's not that this pumping mechanism is battery-free as much as the liquid already *carries* a "battery" on its surface in the form of surface tension.

I find a dance analogy with capillary action helpful when thinking about the phenomenon. When fluid enters a small vertical conduit, capillary action results from the interplay—a microscopic dance, really—between the fluid's surface tension and the wettability of the conduit walls. Gravity (pulling down) and Viscosity (resisting movement) also exert forces. If we could visualize these actors on a vertical capillary as human dancers, we would see the debonair Wetting leading the engaging Surface Tension upward on the vertical dance floor; however, jealous Gravity is continuously dragging Wetting down until Wetting becomes exhausted, and Surface Tension, left without a partner, decides to call it quits. In a horizontal conduit, dancer Gravity is absent, so dancers Wetting and Surface Tension can go uninterrupted for much longer on their horizontal dance floor. But the dance in the horizontal direction is not infinitely long either because, at some point, Viscosity starts resisting the motion of the fluid, and dancer Wetting cannot drag any more liquid and calls it quits as well.

This dance is critical for the fertilization of soil, the transpiration of plants, the chemical communication between animals, and our temperature regulation through sweating, among other phenomena. It has also been the basis for many devices—from the candlewick and the bandage to the ballpoint pen and the COVID test—that have touched most people's lives.

7

The Plants of Progress

Capillary Action in Plants, the Drop-by-Drop Extraction of Useful Juices from Certain Trees, and Autonomous Capillary Microfluidic Devices

On July 4, 1776, as he was about to solemnly sign the Declaration of Independence in the company of great men, Dr. Benjamin Rush demonstrated the seeds of the courageous stand he would show against slavery years later using a by-product of tree microfluidics, maple syrup. The colonies had been at war with Great Britain for over a year. On that fateful day, someone next to him joked about the great perils they would face if the Revolution failed.[1] Rush represented the predominantly loyalist Pennsylvania and was arguably the most famous doctor in the rebel colonies, constantly advising more prominent revolutionaries such as George Washington.[2] He would be one of the first to hang. But his long-time friends and fellow revolutionaries, Benjamin Franklin, John Adams, and Thomas Jefferson, were at his side, so he braved the moment and signed.

After the Revolution, Rush continued his pleasant life as a well-educated Enlightenment intellectual—he was fluent in French, Italian, and Spanish—and a professor of chemistry at the College of Philadelphia (the current University of Pennsylvania). Rush, one of the unsung Founding Fathers, used his wisdom and influence—he befriended the first three presidents—to advocate many reforms, such as free public schools, women's education, the derogation of the death penalty, and humane treatment of the mentally ill. He was also a fierce abolitionist, and at one point, he became convinced that slavery, which pro-

---

[1] Signing of the United States Declaration of Independence. *Wikipedia.* https://en.wikipedia.org/wiki/Signing_of_the_United_States_Declaration_of_Independence (n.d.).

[2] Fried, S. *Rush: Revolution, Madness, and Benjamin Rush, the Visionary Doctor Who Became a Founding Father.* Crown, 2018.

vided free labor for cane sugar plantations,[a] could be financially defeated with a new weapon derived from a microfluidic feature of plants: maple sugaring, the drop-by-drop extraction of sugar from maple trees.[3]

We have a lot to learn about capillarity from Nature. Plants contain many juices that are transported through systems of tiny capillaries. The *sap* is the fluid containing minerals and water that feeds every cell in the plant. The *xylem* is the ensemble of natural tubes in a plant that carries sap up from the roots to the rest of the plant (Figure 7.1, top and middle). The xylem vessels run parallel to the stem of the plant, although they are not continuous like a straw—they are formed of empty dead cells called *tracheids*, each interconnected by small holes or pits that function as valves. The water-loving tracheid walls aid the upward capillary flow by offering more surfaces for the sap to cling on in their climbing journey than a straight, smooth vessel. Thus, contrary to popular belief, the sap ascent through the xylem resembles more a game of Tetris than your sipping of an iced drink through a straw. The diameter of the vessels at the base of the plant can range from 20 micrometers for short plants to 200 micrometers for very tall trees. At the top of the plant, the vessel diameters can narrow by half or more.[4]

There is also a parallel set of vascular structures called the *phloem* that carry liquid nutrients in both directions, coming down from the upper leaves as well as up from the lower leaves (Figure 7.1, top and middle). The phloem is responsible for doling out the products generated during *photosynthesis*. This process is a cascade of light-triggered chemical reactions that occur in the leaves, converting carbon dioxide and water into oxygen and simple sugars, such as glucose and fructose. These simple sugars are then combined into more complex sugars, such as sucrose and starch, for transport and storage. Inside their leaves, stems, stalks, and trunks, plants harbor a marvelous system of parallel microfluidic elevators that shuttle fluids up and down with tireless precision.

---

[a] Truth be told, Washington, Jefferson, and Rush owned slaves as servants in their homes, which seems to contradict the movement of equality for all Americans they were leading. To help judge them with a period-appropriate lens, the slave work in the sugar cane plantations (which they were trying to eradicate) was, by contrast with domestic service, intolerably cruel and dangerous. The East India Company calculated that every 450 pounds of sugar produced (a year-and-a-half consumption of sugar by one family) cost one slave life.[5] For a balanced discussion on the contradictions of these Founding Fathers and slaveholders, see Ambrose.[6]

[3] Theobald, M. M. Thomas Jefferson and the maple sugar scheme. Colonial Williamsburg J. (2012, Autumn). https://research.colonialwilliamsburg.org/Foundation/journal/autumn12/maplesugar.cfm.

[4] Rosell, J. A., Olson, M. E., & Anfodillo, T. Scaling of xylem vessel diameter with plant size: Causes, predictions, and outstanding questions. *Curr. For. Reports* 3: 46–59 (2017).

[5] Pearsall, G. Maple syrup production and slavery. *Adirondack Almanack*. https://www.adirondackalmanack.com/2015/03/maple-syrup-production-and-slavery-2.html (2015, March 25).

[6] Ambrose, S. E. Founding fathers and slaveholders. *Smithsonian Magazine*. https://www.smithsonianmag.com/history/founding-fathers-and-slaveholders-72262393 (2002, November).

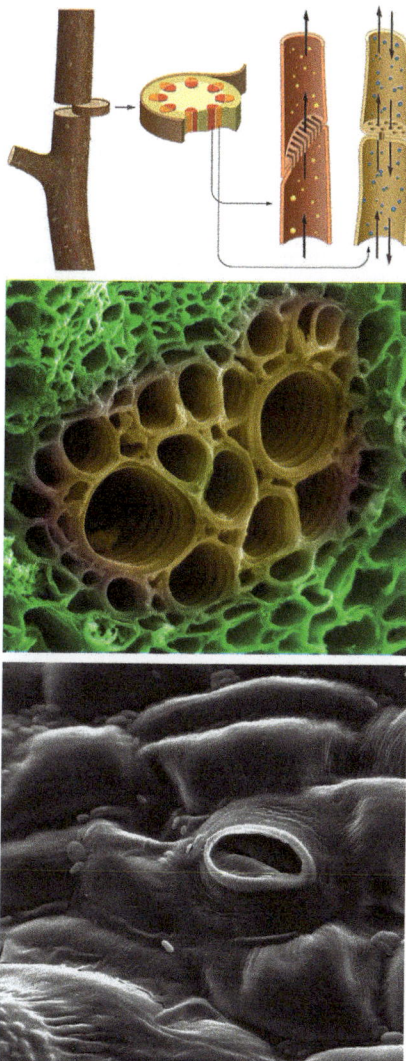

Figure 7.1 (Top) Schematic of the structure and transport mechanism in plants, showing the xylem's upward flow (left tube) and the phloem's bidirectional flow (right tube). (Middle) Xylem and phloem vascular structures (false colored in yellow) are both visible in a geranium leaf. The diameter of the largest conduits is approximately 30 micrometers. The surrounding green-colored part is formed of the cells where the photosynthesis occurs. This image was taken with a scanning electron microscope. (Bottom) A stoma of a tomato plant leaf. The opening is 5 micrometers wide and 10 micrometers long.

Image sources: (Top) Aldona Griskeviciene/Shutterstock. https://www.shutterstock.com/image-illustration/internal-anatomy-tree-stem-2217134193. (Middle) Courtesy of Professor David Furness, Keele University. (Bottom) User Shizhao on Wikipedia. https://commons.wikimedia.org/wiki/File:Tomato_leaf_stomate_1.jpg. Public domain.

Early French explorers in the 1500s watched various Native North American tribes obtain a sweet, transparent juice by slitting the bark of maple trees.[7] The nations of the Anishinaabe (including the Ojibwe, the Algonquin, and the Mississaugas, among others) and the Haudenosaunee (including the Mohawk, the Onondaga, and the Seneca, among others), all peoples who lived in vast expanses bordering what is now the United States and Canada, were using similar techniques to extract the fluid and process different varieties of sugar from it.[b] A Haudenosaunee legend says that a tomahawk thrown by Chief Woksis at a maple tree produced a cut, and the juice from the incision was used to cook meat, which tasted sweet. Most likely, the natives had tasted the "sapsicles"—icy drippings—that are occasionally observed on the broken twigs of maple trees on freezing days.[8]

The sugar production by maple trees has very tight physiological requirements. The trees store the starch produced during photosynthesis in their trunks and roots during spring and summer. When the trees need to pull from their reserves in late winter and early spring, they convert the starch to sugar, which rises by capillarity in the sap.

How did plants acquire this power to move fluids? Plants are energy savers. They stay in the same place and are autonomous organisms that generate their food through photosynthesis. They do not have power-hungry muscles for locomotion, they do not forage for food, and they did not grow hearts because muscles consume too much energy. Terrestrial plants evolved from algae to live on land approximately 450–500 million years ago.[9] These first land plants did not have leaves, a stem, or roots—they lacked vascular structures and were short, much like a moss. A major event in the evolution of terrestrial plants was the "invention" of a *vasculature* about 400 million years ago[10] for a new type of *vascular plants*,[11] which were very fern-like.[12] It's as if Nature had suddenly "realized" that fluids (water and nutrients) could move against gravity from the base up the plant's structures using capillary wicking. In reality, capillarity in plants did not start by some intelligent design but, rather, by a slow

[b] This *sucre du pays* (country sugar) made it all the way to Versailles, to the court of Louis XIV, the seventeenth-century almighty *Roi Soleil*, who loved it.[7]

[7] Les moment forts de l'érable. *Érable du Québec*. https://erableduquebec.ca/a-propos/histoire (2024).

[8] Massachusetts Maple Producers Association. Maple history. https://www.massmaple.org/about-maple-syrup/maple-history (n.d.).

[9] Delwiche, C. F., & Cooper, E. D. The evolutionary origin of a terrestrial flora. *Curr. Biol.* 25: R899–R910 (2015).

[10] Kenrick, P., & Crane, P. R. The origin and early evolution of plants on land. *Nature* 389: 33–39 (1997).

[11] Sen, A. Tree fern genome provides insights into its evolution. Carl R. Woese Institute for Genomic Biology, University of Illinois at Champaign–Urbana (2022).

[12] Huang, X., Wang, W., Gong, T., Wickell, D., Kuo, L. Y., Zhang, X., et al. The flying spider-monkey tree fern genome provides insights into fern evolution and arborescence. *Nat. Plants* 8: 500–512 (2022).

process of genetic trial-and-error that lasted tens of millions of years. And when this endless genetic roulette finally produced plants with vascular structures, these plants had the advantageous ability of growing taller than their neighbors in search for light, and they passed this trait (aptly called *arborescence* for "tree-shaped") to their offspring. Natural selection has infinite patience: Sooner or later, disadvantageous traits are eliminated through competition, whereas advantageous traits are passed down to the descendants. A key aspect of the new mechanism of capillarity is that it did not cost the plants any energy expenditure because it only utilized the energy stored in the surface tension of water. Several million years later, those early ferns evolved into trees, which genetically inherited this clever microfluidic mechanism and passed it down to their progeny, until they covered the planet with the magnificent forests of the current era. All those timberlands you see are powered by capillarity. Today, more than 90% of Earth's vegetation is formed of vascular plants[13]—just about everything that is not a moss.[c]

But there is a problem: If you do the math (I'll spare you), you will find that capillary action alone is not strong enough to pump water from the roots all the way up to the top of a tall tree. So how do trees, without a beating heart, pump the sap from their roots to their tallest branches? The answer is that they get help from an additional microfluidic trick—at the leaves. Leaves appeared in evolution approximately 360 million years ago,[14] so it is fair to say that in our jest of looking at Nature as a budding microfluidic engineer, she was a bit slow and still trying to figure things out. Not once, but twice in evolution, Nature reached into its microfluidic toolbox to give vascular plants an extra boost to get through natural selection. The xylem branches out and connects with conduits inside the leaves, which end in 5- to 10-micrometer-wide mouth-shaped pores on the surface of the leaves called *stomata* (Figure 7.1, bottom). The density of stomata can vary with individual, species, and environmental conditions, from a few stomata to a thousand stomata per millimeter square.[15] A pair of *guard cells* flank each stoma to regulate the pore opening. The stomata open during the day to let carbon dioxide into the leaf (necessary for photosynthesis) and close at night to save water (by limiting evaporation).[16] When the stomata are open during the day, the sap in the xylem evaporates through the pores. Evaporation has the same effect as your suction when you drink with a straw. (It also helps

---

[c] Also included in the group of nonvascular plants are liverworts and hornworts.
[13] Lumen Learning. Seedless vascular plants. SUNY–OER Services. https://courses.lumenlearning.com/wm-biology2/chapter/seedless-vascular-plants (n.d.).
[14] Beerling, D. J., Osborne, C. P., & Chaloner, W. G. Evolution of leaf-form in land plants linked to atmospheric CO_2 decline in the Late Palaeozoic era. *Nature* 410: 352–354 (2001).
[15] Tay, A. C., & Furukawa, A. Variations in leaf stomatal density and distribution of 53 vine species in Japan. *Plant Species Biol.* 23: 2–8 (2008).
[16] Costa, J. M., Monnet, F., Jannaud, D., Leonhardt, N., Ksas, B., Reiter, I. M., et al. Open all night long: The dark side of stomatal control. *Plant Physiol.* 167: 289–294 (2015).

that the sap likes to crawl up the walls because the xylem is very hydrophilic, and the sap has a high surface tension, so it likes to stay together.) Scientists call this evaporation-aided, capillary-based microfluidic pumping mechanism to bring fluid up the body of plants *transpiration*.

At the pores of leaves, our dance metaphor would look like an eternal rehearsal: Choreographer Evaporation continuously encourages the dancers by removing a bit of water, pulling dancer Wetting up a bit, which makes dancer Surface Tension lead again for an instant until dancer Wetting comes back in full swing. Ever-jealous Gravity wants to end the dance. When it gets too hot, some plants can shut off the dance altogether by closing their leaves' stomata to save water.[d] As precise as a clock and common to all leafed plants, this microscale fluidic dance is so exact and reliable that these organisms have evolved to depend on it.

All the leaves do the pumping in parallel, resulting in huge volumes of water being taken from the ground and (almost all of it) evaporating through the leaves. This water vapor goes back to the atmosphere and helps make rain. An acre of corn can transpire between 500 and 600 liters of water every hour, and a large oak tree can transpire more than 400 liters a day.[17] By comparison, humans drink an average of 2.5–3.5 liters a day. The lives of plants are powered by millions of microfluidic pumps (the stomata) working in unison on each leaf, each pump connecting to a capillary conduit of the xylem, precisely to recycle the water that sustains them. Nature uses microfluidics to maintain its vast gardens.

The North American tribes had rudimentary knowledge about these pumping rates and had learned to use them for maple sugaring. They observed that sap starts flowing when temperatures alternate between freezing at night and thawing during the day.[e] They learned to slash the bark during the "sugar month" and funnel large quantities of sap with a wedge, drop by drop, into bark baskets (Figure 7.2). The clear liquid was then concentrated to brownish, grainy sugar

---

[d] This microscale mechanism also illuminates the fragility of life on Earth. Inside the leaves, photosynthesis uses carbon dioxide to build chemical compounds such as sugars that feed the plant through the sap. A by-product of this chemical reaction is oxygen, which is released by the plant at the leaves. Approximately one-third of the oxygen on Earth is produced by terrestrial plants and the other two-thirds by marine plants. Photosynthesis cannot take place in most plants above a temperature of around 47°C—a plausible scenario in a global warming future: In July 2023, officially the world's hottest year on record, a billboard in Phoenix, Arizona, recorded 47.8°C. When photosynthesis stops, the plant dies. If all plants died, the oxygen on Earth would not be renewed, and only organisms such as certain bacteria that live without oxygen would be able to survive.

[e] European maple trees do not undergo these sharp temperature transitions needed for sugar to flow in the sap. As a result, the ancient peoples of Europe did not develop maple sugaring.

[17] Water Science School. Evapotranspiration and the water cycle. U.S. Geological Survey. https://www.usgs.gov/special-topics/water-science-school/science/evapotranspiration-and-water-cycle?qt-science_center_objects=0#qt-science_center_objects (2018).

Figure 7.2 (Left) Ojibwe woman tapping a sugar maple in 1908. (Right) Detail of a modern setup for tapping a maple tree.

Image sources: (Left) U.S. Library of Congress, photograph by Roland Reed. https://loc.gov/pictures/resource/cph.3c05740/. (Right) Luce Morin/Shutterstock. https://www.shutterstock.com/image-photo/maple-sap-dripping-into-aluminium-bucket-2238949785.

using a combination of freezing (the sugary bottom does not freeze, and the frozen top is discarded) and heat (the water evaporates, leaving sugar).[18]

It was a lot of work.[f] In 1846, a Jesuit wrote,

> At the beginning, our people are in the snow to collect the sap that flows *drop by drop* from the notches they have made in the trees. Below these openings they place small bark cups they have in large numbers and are always on the run to see if these cups are full and then pour them into large tubs. Once this sap is collected, it must be brought to a boil; practically all of it is dissipated in steam. A barrel of sap yields perhaps one pound of sugar.[19]

The British colonists imitated the natives' process and improved its yield with better materials and a drilled hole or "tap" (Figure 7.2, right). They realized

---

[f] As a rule of thumb, 40 liters of sap yield 1 liter of maple syrup.[20]

[18] Nearing, H., & Nearing, S. *The Maple Sugar Book: Together with Remarks on Pioneering as a Way of Living in the Twentieth Century.* Schocken, 1971.

[19] Corbiere, A. Ninaatigwaaboo (maple tree water): An Anishinaabe history of maple sugaring. The Great Lakes Research Alliance. https://grasac.artsci.utoronto.ca/?p=136 (2015).

[20] Bruchac, J. How making maple syrup keeps native culture alive. Literary Hub. https://lithub.com/how-making-maple-syrup-keeps-native-culture-alive (2022, September 2).

maple sugar was much simpler—and thus less expensive—to produce than the cane sugar imported from the Caribbean and saw in maple sugaring a commercial opportunity in the form of extra income for farmers.[g] Rush and other abolitionists saw their chance "to lessen or destroy the consumption of West Indian sugar, and thus indirectly to destroy negro slavery"[21] and started recruiting fellow revolutionaries to their cause. Jefferson was one of them. In 1789, two years before becoming the third president, Jefferson joined Rush's Society for Promoting the Manufacture of Sugar from the Sugar Maple Tree and started a plantation of maple tree saplings in Monticello, Virginia. Alas, Jefferson's saplings didn't grow well in Monticello. However, many farmers in New England started sugaring operations, and by 1818, maple sugar was selling for half the price of imported cane sugar.[8]

Rush's scheme only worked briefly (speculation started a "maple bubble" that ultimately burst). Still, slavery's days were numbered anyway for reasons that went beyond its financial soundness. Britain passed the Slavery Abolition Act in 1834, and the United States followed suit in 1865. By 1880, cane and maple sugar were approximately equal in price,[8] but it didn't matter anymore. The maple sugaring industry quickly repurposed itself to the production of maple syrup, and it rapidly became the favorite add-on to pancakes, waffles, and French toast for breakfast in America and Canada.

* * *

Another gift to human civilization derived from tapping the sap of plants also arose from the Natives of the Americas. As far as we know, it started with the Olmecs, who lived in the isthmus of Tehuantepec, irrigated by three tropical rivers—the Papaloapan, the Coatzacoalcos, and the Tonalá—in the plains of modern-day southern Mexican states of Tabasco and Veracruz. Here, the Olmecs rose to a prosperous society of corn-growing farmers, fine sculptors, and pioneering merchants that flourished for over a millennium, starting approximately 3,600 years ago (Figure 7.3).[h] They built some of the continent's first urban centers and the first pyramid. They traveled long distances to trade their jade or obsidian jewelry, pottery, drinkable chocolate, polished mirrors, and

---

[g] Maple sugaring was popular among farmers because it was done during the winter when other farming activities were at their lowest. A single family could make as much as 1,000 pounds of sugar per season.[8]

[h] The Olmec civilization declined around 400 BCE,[22] but the legacy they left behind through their trade exchanges remained.

[i] "Olmec" derives from the word "Omecatl" used in Nahuatl (the language of the Aztecs), composed of "olli"—meaning "natural rubber"—and "mecatl"—meaning "people."

[21] Davis, K. C. A sweet assault on slavery. *Huffpost.* https://www.huffpost.com/entry/maple-sugar-slavery_b_833972 (2011, March 17).

[22] Minster, C. The decline of the Olmec civilization: The fall of the first Mesoamerican culture. ThoughtCo. https://www.thoughtco.com/the-decline-of-the-olmec-civilization-2136291 (2018).

Figure 7.3 Olmec head in Villahermosa, capital of Tabasco, Mexico.

Image source: Pxhere.com. https://pxhere.com/en/photo/1235440.

Figure 7.4 Latex being collected from a tapped rubber tree.

Image source: User Ji-Elle on Wikipedia. https://en.m.wikipedia.org/wiki/File:Sri_Lanka-Rubber_plantation_%285%29.JPG (CC BY-SA 3.0).

rubber objects. Because they taught others how to make rubber, they became known throughout Mesoamerica as Olmecs or "the rubber people." [i,23]

The Olmecs had observed that they could extract *latex* from the native Panama rubber trees (*Castilla elastica*) by tapping or making an incision in the bark (Figure 7.4). They did the work at night or in the early morning before the day's heat caused the latex to coagulate and seal the cut. The Olmecs used rubber to make shoe soles, elastic bands, hollow figurines, and balls. They also noticed the unique water impermeability of rubber and had soaked clothes with it to protect themselves from the rain.

---

[23] Cartwright, M. Olmec civilization. *World History Encyclopedia.* https://www.worldhistory.org/Olmec_Civilization (2018).

In addition to the sap, plants have other juices—such as the latex of rubber trees and the resin of pine trees—they use as a defense mechanism.[j] Latex runs through capillaries that are about 25 micrometers in diameter. The latex of rubber trees consists of an aqueous suspension of particles about half a micrometer in diameter made of very long strands of a linear-shaped rubbery polymer called *cis*-polyisoprene. The mix contains approximately 70% water and 39% rubber.[k]

Drop by drop, the latex extracted from the rubber tree changed the customs and destinies of entire nations.[l] With it, the Olmecs developed the first rubber balls and a ballgame that the Mayas and then the Aztecs later adopted.[m] Indigenous people of Mexico still play a modern version of the game called *ulama*, making the Mesoamerican ballgame the oldest sport in history and the first sport ever to use a rubber ball (Figure 7.5). Ulama resembles a netless volleyball and was originally played with armor for protection against the heavy 10-pound ball. Estimate how many rubber-based balls you might have owned, and you will realize how different your life would have been without that Olmec heritage.

This Olmec invention derived from tapping a plant would have, in time, a planetary impact much more significant than the gold that the Spanish invaders were so interested in. In 1528, Hernán Cortés sent a troupe of *ōllamanime* (ballplayers) to Spain to perform for King Charles V.[26] The properties of rubber fascinated Europeans. Still, they did not reach past academic circles for a couple of centuries until the start of the Industrial Revolution, when engineers applied the elastic material to mechanical and electrical inventions. By chance, French explorer and geographer Charles Marie de la Condamine traveled to Ecuador in 1736 to confirm a prediction by Newton that the Earth bulges in diameter at the equator due to gravitational effects. It was in Ecuador that De la Condamine saw latex for the first time.

---

[j] Latex is a generic name that refers to a milky juice secreted by a network of cells called *laticifers* located in the outer layers of certain plants. As many as 14% of tropical plants produce latex. Latex contains toxins as a defense mechanism against plant-eating insects and also coagulates at warm temperatures to seal a wound. The white fluid inside the stalk of a dandelion is also latex but does not contain enough rubber to be commercially useful.[24] Other plants have similar defense mechanisms—for example, pine trees ooze out *resin* through their *resin canals* when their bark is punctured.

[k] Not all latex contains rubber. Opium, for example, is the dried form of the latex extracted by slashing the unripe seedpods of the *Papaver somniferum* poppy.

[l] Aside from the major agricultural crops (potato, corn, wheat, and rice), only a handful of plants have significantly influenced human economies: the rubber tree, the wine grape, the tobacco tree, the coffee plant, and timber. See Hobhouse.[25]

[m] The Mayas flourished in the Yucatán Peninsula, east of the Olmec territory, between 2,000 BCE and the sixteenth century when they came into contact with the Spanish. The Aztecs, on the other hand, established their empire west of the Olmec territory that lasted from 1,300 AD until they were decimated by the Spanish.

[24] SPC. History of rubber. https://www.spc-group.com/about-rubber/history-of-rubber (n.d.).

[25] de la Garza, M., & Izquierdo, A. L. El Ullamaliztli en el siglo XVI. *Estud. Cult. Nahuatl* 14: 315–333 (2022). https://nahuatl.historicas.unam.mx/index.php/ecn/article/view/78430/69380

[26] Hobhouse, H. *Seeds of Wealth: Five Plants That Made Men Rich*. Shoemaker & Hoard, 2003, 2005.

Figure 7.5 Modern ulama player in Sinaloa, Mexico.

Image source: *User Manuel Aguilar on Wikipedia. https://en.m.wikipedia.org/wiki/*
File:Ulama_37_%28Aguilar%29.jpg (CC BY 2.5).

De la Condamine had brought botanist François Fresneau with him. When
the two ran into a tribe of Tsachali people that were tapping latex, they were
intrigued. The Quechua-speaking natives called it "cahuchu," or "weeping
wood."[27] Back in France, in 1736, De la Condamine presented rubber samples to
the Académie Royale des Sciences. In 1755, Fresneau published the first scientific paper on rubber, describing its properties. British chemist Joseph Priestley,
the discoverer of oxygen and inventor of carbonated water, must have been
enthralled by this unique material when he agreed to pay three shillings for a
half-inch cube at a London shop on April 15, 1770.[n] He noticed he could use it to
rub off pencil marks.[24] The word "rubber" quickly stuck among the Royal Society Fellows, who must have found the word "caoutchouc"—the French version

[n] A hefty sum at today's money of $20 for a 2-gram piece. That means Priestley paid $10,000/kg
for rubber.
[27] Polymer Science Learning Center. Ecuador: Sometimes useful. https://pslc.ws/macrog/exp/
rubber/bepisode/euro.htm (2002).

of "cahuchu"—too challenging to pronounce.° This rubber became our beloved chewing gum and essential pencil eraser.

The rubber industry exploded in the West in the 1800s, coinciding with the combined revolutions in industrial mechanization and modern chemistry. The first industrial rubber production plant was established in Paris in 1803. The company made rubber bands used for braces and garter suspenders. In 1824, Scottish chemist Charles Macintosh invented a waterproof raincoat by cementing a rubber layer between two layers of cotton fabric.[P, 28] But natural rubber was not perfect back then—it was rather sticky at room temperature, became stickier on hot days, and was hard and stiff on cold days. American inventor Charles Goodyear heated the rubber with vapors of sulfur and lead—a process he patented in 1844 and termed *vulcanization* after Vulcan, the Roman god of fire—to make rubber more stable with varying temperature.[29]

Almost innumerable uses for rubber followed. In 1846, Scottish engineer Robert William Thomson patented a pneumatic rubber tire consisting of a hollow belt of rubber inflated with air so that the wheels presented "a cushion of air to the ground, rail or track on which they run."[24] A book published in 1857 by British inventor Thomas Hancock—co-founder of Charles Macintosh Corporation—described hundreds of commercial inventions spanning inflatable cushions; valves; gaskets; tubing; tires; boats; rafts; diving suits; buoys; rainproof coats and boots; hospital articles such as a neck or pillow and a water bed; domestic objects such as rubber bottle corks, bathing caps, and baby pacifiers; and sporting goods such as waterproof tents, canoes, and—of course—rubber balls.[31] Without the tapping technique—which extracts drop after drop of latex from the trees' capillaries—there could have never been a rubber craze of global proportions in the nineteenth century.

By then, manufacturers had transferred most of the rubber production to the Amazon, where the superior variety of rubber trees, *Hevea brasiliensis*, grew in abundance. The *H. brasiliensis* is 50% taller than the *C. elastica* exploited by the Olmecs, and its latex vessels are much more interconnected. Thus, when the *H. brasiliensis* trees are tapped, the latex flows more quickly down the bark.

° Spanish speakers still use "caucho" to designate natural rubber.

P The process—which started the Mackintosh brand of impermeable garments and later inspired the punk latex fashion in the 1970s[30]—essentially rubberized the cloth by dissolving the rubber in a solvent. The spelling in the "Mackintosh" brand (by the Macintosh Corporation) was intentional.

[28] Mackintosh. Brand story. https://www.mackintosh.com/us/brand-story (n.d.).

[29] Goodyear, C. Improvement in India-rubber fabrics. U.S. Patent Office. https://en.wikisource.org/wiki/United_States_patent_3633 (1844).

[30] George, C. From fetish to fashion: The rise of latex. BBC.com. https://www.bbc.com/culture/article/20200108-from-fetish-to-fashion-the-rise-of-latex (2020, January 8).

[31] Hancock, T. *Personal Narrative of the Origin and Progress of the Caoutchouc or India-Rubber Manufacture in England*. Longman, Brown, Green, Longmans & Roberts, 1857.

The amount of latex harvested from each tree per day was critical because firms such as the Macintosh Corporation sought a lot of rubber, and the dripping method of latex extraction was stifling their ambitions. There was an "Amazon rubber boom" in the second half of the nineteenth century, with people traveling to Brazil to profit from the extraction of this "white gold" and towns swelling with immigrants.

Lacking the proper international commerce protections, Brazil's monopoly was effortlessly bypassed by more powerful nations. In 1876, British business-man Henry Alexander Wickham smuggled 70,000 *H. brasiliensis* seeds wrapped inside banana leaves and shipped them to the Royal Botanical Gardens in London. The surviving 1,900 seedlings were sent to Malaysia and from there to India, Sri Lanka, and Indonesia, where the British established large rubber tree plantations despite Brazil's protests. The British fraud overcame the slow, handmade production based on capillary dripping from Mesoamerican rubber trees: Millions were planted in Asia and Africa, where labor could be readily exploited.

Until the beginning of the twentieth century, almost all Western civilization depended on Olmec-style rubber extraction to make balls, tires, tubes, gaskets, and various apparel, resulting in the colonial exploitation of rubber. The human rights abuses were horrific.[q] The Asian and African rubber tree plantations deflated the Amazon rubber boom. They gave the British Empire a de facto monopoly on natural rubber at the beginning of the twentieth century until synthetic rubber displaced natural rubber.[34] To this day, however, synthetic rubber cannot match latex's superior flexibility, electrical insulation, and corrosion resistance. We still depend on natural rubber to make essential medical

---

[q] The most famous example was denounced by Black historian George Washington Williams in 1890 as a "crime against humanity"—the first such accusation in history—against King Leopold II of Belgium for personally supporting the atrocities committed by the Anglo-Belgian India Rubber (ABIR) company.[32] (Leopold II was the owner and absolute ruler of the Congo at the time.) In 1906, French-born British journalist, socialist politician, and pacifist Edmund D. Morel wrote the accusatory book *Red Rubber* to unveil the slave-supported rubber industry in the Congo.[33] His book, and the photographs of Christian missionaries such as Alice Harris and her husband detailing amputations and torture of natives by sentries of the ABIR, contributed to the success of a campaign—with the support of writers such as Sir Arthur Conan Doyle and Mark Twain—that ended Leopold II's personal ownership of Congo. The anti-slavery Congo campaign led by Morel and initiated by Williams is considered the first human rights campaign in history and still stands as an inspiration to other human rights organizations and movements.[35]

[32] Williams, G. W. Letter to the American Secretary of State [1890]. In *Aux Origines de L'État Indepéndeant du Congo* (ed. Bontinck, F.). L'Universite Lovanium de Leopoldville, 1966.

[33] Morel, E. D. *Red Rubber: The Story of the Rubber Slave Trade Which Flourished on the Congo for Twenty Years, 1890–1910.* National Labour Press, 1906.

[34] Arias, M., & van Dijk, P. J. What is natural rubber and why are we searching for new sources? *Frontiers for Young Minds.* https://kids.frontiersin.org/articles/10.3389/frym.2019.00100 (2019, July 19).

[35] Sliwinski, S. The Kodak on the Congo: The childhood of human rights. Autograph ABP (2010).

devices, surgical gloves, aircraft tires, pacifiers, toys, and clothes.[36] Complicitly, we have all become *rubber people.*

There is abundant evidence that the Indigenous civilizations of Mesoamerica had learned to appreciate many juices circulating within plants other than latex. They were botanists and knew about rainforest plants that could neutralize a snake's poison or reduce a headache.[37] The Olmecs learned to process the milky sap into elastic, bouncy caoutchouc by adding the juice from morning glory vines that grew near these trees and heating the mixture.[38,r] The Olmecs must have been familiar with these vines and their hallucinatory properties,[39] as they used them in rituals. When the conquistadores led by Hernán Cortés arrived in 1519 at Tenochtitlán, the Aztec capital built on an island in the middle of Lake Texcoco, they were astonished at all the marvelous botanical gardens they saw.[40] The Chapultepec and the Oaxtepec gardens, near Tenochtitlán, had irrigation systems and were bursting with medicinal herbs and edible plants—the "green gold" the Spanish took back to the Old World.[s]

Plants are foreign to human conquests and will grow ignoring our maple syrup breakfasts or rubber ball games. Barring catastrophic climate change, they will continue their transpiration dance for centuries, allowing trees to outgrow and outlive any other living organism on Earth. Sugar maples can live up to 400 years, so the maple syrup you enjoyed on your last pancake might have come from a tree that was already well-grown when Dr. Benjamin Rush and his fellow revolutionaries signed the U.S. Declaration of Independence. But maple

[r] The enhanced elastic behavior of the rubber relative to the unprocessed latex was likely due to organic compounds present in the species of morning glory vine *Ipomoea alba* used by the Olmecs. These organic compounds appear to have induced the purification of the polymer component in the latex and an increase in the strength and number of interchain interactions in the rubber.[38] This process is conceptually similar to the stabilization of rubber by vulcanization developed by Charles Goodyear in 1844, three millennia later.[29]

[s] The Spanish invaders were very ungrateful for this health-related knowledge the "Indians" gave them. Cortés and his henchmen looted and flattened Tenochtitlán in 1521 and spread smallpox among the natives—unintentionally, but with no signs of remorse: They enslavened the ones left to build Mexico City over the ruins of their homes. At Oaxtepec, they built the Hospital de Santa Cruz, to which the natives had no access.[41]

[36] Vijayaram, T. R. A technical review on rubber. *Int. J. Des. Manuf. Tech.* 3: 25–36 (2009).

[37] Thun, H. El saber médico de los guaraníes y la medicina de los jesuitas. In *El Conocimiento Indígena como Recurso: Transmisión, Recepción e Interacción del Conocimiento entre América y Europa, 1492–1800* (eds. Dierksmeier, L., Fechner, F., & Takeda, K.). Tubingen University Press, 2021.

[38] Hosler, D., Burkett, S. L. & Tarkanian, M. J. Prehistoric polymers: Rubber processing in ancient Mesoamerica. *Science* 284: 1988–1991 (1999).

[39] Beaulieu, W. T., Panaccione, D. G., Quach, Q. N., Smoot, K. L., & Clay, K. Diversification of ergot alkaloids and heritable fungal symbionts in morning glories. *Commun. Biol.* 4: 1–11 (2021).

[40] Castillo, B. D. del. *La Verdadera Historia de la Conquista de la Nueva España* (1576).

[41] Yáñez, R. G. Exhospital de la Santa Cruz. Instituto Nacional de Antropología e Historia. https://mediateca.inah.gob.mx/repositorio/islandora/object/guia%3A234 (1996).

trees are not even the oldest and tallest, primarily concentrated in California and the Pacific Northwest.[t]

The tallest tree is a 116-meter-tall sequoia (also commonly termed "coast redwood") named Hyperion in northern California that has been growing for

Figure 7.6 The General Sherman tree in Sequoia National Park, California.

Image source: User Tuxyso on Wikipedia. https://en.m.wikipedia.org/wiki/ File:General_Sherman_Tree_2013.jpg (CC BY-SA 3.0).

[t] World-class giants such as the Douglas fir, Sitka spruce, noble fir, western hemlock, ponderosa pine, and grand fir are all native to Washington state.[42]

[42] McCrate, C. Science offers compelling theories for the mysteries of our tallest trees, but their majesty requires no research—just appreciation. *Seattle Times*. https://www.seattletimes.com/ pacific-nw-magazine/sept-20-grow (2020, September 19).

almost 800 years. The General Sherman—the largest tree in trunk volume—is a magnificent 84-meter-tall, 31-meter-circumference giant sequoia that can be admired by conveniently driving or strolling through Sequoia National Park (Figure 7.6). The oldest known tree is a 4,853-year-old bristlecone pine named Methuselah in the White Mountains of eastern California.[u] Picture this: Methuselah was born before the Egyptians built the Pyramids of Giza and was already well-grown at 3,000 years old when the Olmecs started growing corn and tapping latex in the Gulf of Mexico.

The microfluidic transpiration mechanism can lift almost 2,000 liters of water up the trunk of a sequoia every day.[43] Measurements of the sap flow rate in sequoias at different months of the year[44] tell us how these magnificent plants deal with their own size. In fact, they have grown to the limits of their possibilities[45]—if they grew a bit more, the top branches would die because the pumping mechanism would fail, for various reasons. First, like humans, they can suffer "embolisms"—as the height increases, the pressure of the fluid column pulling down can suck air into a tracheid, causing the formation of a bubble that interrupts the sap flow. Also, all the tallest trees are evergreens. They cannot afford to lose their leaves seasonally because the sap would barely reach the top in time to grow new leaves in the spring. In the coldest and wettest month (December), the sap flow rate is reduced to 5 meters per month; at this time of the year, the leaves' stomata are often covered by water—which in the cold does not evaporate well, so transpiration is not very effective—and the cold sap is more viscous. The leaves do not get much sap in the coldest months, which is just as well because the plant consumes very little energy during the winter. On the other hand, in the hottest and driest months (June and July), the leaves are dry, evaporation goes at full steam, and the sap is runnier; the sap ascends at the highest rate of 50 meters per month, ten times faster than in the winter.[44] These are the months when photosynthesis and transpiration are most active, and the tree grows.

The giant sequoia, General Sherman, did not stop growing throughout much of our history. If it were felled and placed to float on the sea without its roots, it would be three or four times longer than an adult blue whale, the largest animal on Earth. General Sherman is estimated to be between 2,200 and 2,700 years old.

---

[u] The exact location of Hyperion and Methuselah is kept secret because the National Parks custodians have learned better than to trust fellow humans.

[43] Eureka. Sequoia quiz. https://www.pbs.org (n.d.).

[44] Burgess, S. S. O., & Dawson, T. E. The contribution of fog to the water relations of Sequoia sempervirens (D. Don): Foliar uptake and prevention of dehydration. *Plant. Cell Environ.* 27: 1023–1034 (2004).

[45] Koch, G. W., Stillet, S. C., Jennings, G. M., & Davis, S. D. The limits to tree height. *Nature* 428: 851–854 (2004).

[44] Burgess, S. S. O., & Dawson, T. E. The contribution of fog to the water relations of Sequoia sempervirens (D. Don): Foliar uptake and prevention of dehydration. *Plant. Cell Environ.* 27: 1023–1034 (2004).

It was a little sprout when the great Greek philosophers Democritus, Aristotle, Socrates, and Plato lived, sometime around the third and fifth centuries BC—just about the time when the Olmec civilization disappeared while they passed their botanical secrets onto future generations. These trees are a marvelous display of the robustness of life, supported by microfluidics.

* * *

It is no surprise, then, that the capillary action that moves fluids in plants has generated as much fascination as design insights in the minds of countless microfluidic engineers. "Trees were an important inspiration—they pump water, and then they stop. Trees helped conceptualize the idea of autonomous capillary systems that are both self-powered and self-regulated," says David Juncker, Professor of Biomedical Engineering at McGill University in Montreal, Canada.[46] Born in Switzerland and now also Canadian, the tall maple trees in his backyard are a constant reminder of the vision he started developing during his PhD thesis at IBM Zurich under the supervision of Emmanuel Delamarche: the control and confinement of fluids by capillary effects. Throughout the years, Delamarche and his army of microfluidic engineers developed capillary[47,48] and evaporation[49] pumps; pinning valves[50]; and "phaseguides"[51] that, not unlike the tracheids and the transpiration of plants, help liquids fill and find their way through microchannels.

Later at McGill, Juncker and his group elaborated the concept of preprogrammed, autonomous capillary systems—which they termed *capillarics*,[52,53] a fusion of "capillary" and "electronics." Much like electrical engineers build complex circuits like amplifiers from simple elements such as transistors, capacitors, and resistors, Juncker's group designs self-powered microfluidic circuits in modules that include capillary pumps, trigger valves, retention valves, retention burst valves, flow resistors, inlets, and vents.

[46] Folch, A. *Hidden in Plain Sight: The History, Science, and Engineering of Microfluidic Technology.* MIT Press, 2022, p. 123.

[47] Juncker, D., Schmid, H., Drechsler, U., Wolf, H., Wolf, M., Michel, B., et al. Autonomous microfluidic capillary system. *Anal. Chem.* 74: 6139–6144 (2002).

[48] Zimmermann, M., Hunziker, P., & Delamarche, E. Capillary pumps for autonomous capillary systems. *Lab Chip* 7: 119–125 (2007).

[49] Zimmermann, M., Bentley, S., Schmid, H., Hunziker, P., & Delamarche, E. Continuous flow in open microfluidics using controlled evaporation. *Lab Chip* 5: 1355–1359 (2005).

[50] Arango, Y., Temiz, Y., Gökçe, O., & Delamarche, E. Electro-actuated valves and self-vented channels enable programmable flow control and monitoring in capillary-driven microfluidics. *Sci. Adv.* 6: eaay8305 (2020).

[51] Vulto, P., Podszun, S., Meyer, P., Hermann, C., Manz, A., & Urban, G. A. Phaseguides: A paradigm shift in microfluidic priming and emptying. *Lab Chip* 11: 1596–1602 (2011).

[52] Safavieh, R., & Juncker, D. Capillarics: Pre-programmed, self-powered microfluidic circuits built from capillary elements. *Lab Chip* 13: 4180–4189 (2013).

[53] Olanrewaju, A., Beaugrand, M., Yafia, M., & Juncker, D. Capillary microfluidics in microchannels: From microfluidic networks to capillaric circuits. *Lab Chip* 18: 2323–2347 (2018).

Figure 7.7 An example of a capillarics immunoassay in action. The whole movie lasts about 48 minutes. In the first three frames, the user loads the three reagents, the wash buffer, and the sample, which are colored here with dyes for illustration purposes. The sponge layers at the left of the device act as a "pump." The red, yellow, and green reagents are an antibody, an enzyme, and an enzyme substrate, respectively, used to detect the presence of a protein in the yellow sample added in the third frame. As soon as the user fills the last reservoir in the third frame, the sponge starts pumping and the assay starts. The chip automatically aliquots the reagents and sequentially delivers them following a preprogrammed timing without further user intervention. The blue liquid is used for washes in between steps. All the chemical reactions occur on the bottom left, arrow-looking membrane strip connected to the sponge. In the absence of the dyes, the results can be visualized by eye as the appearance of a stripe.

Images courtesy of Houda Shafique and David Juncker, McGill University.

To build valves without a mechanical component, they use surface tension. For example, they design one of these valves as a simple opening at the end of a channel that forms a T-junction with another channel. When the fluid arrives at the junction, it stops because the surface tension is too high to overcome the small aperture. However, if more fluid is added, the valve bursts—similar to the pit of a tracheid during sap ascent—so the assay may continue. "I vividly remember my exhilaration at the time of the invention!" recalls Juncker from his days in the Delamarche lab when he first came up with this retention burst valve.[46]

Using this trick and others, they have demonstrated a microfluidic circuit that measures the protein concentration in the blood (called C-reactive protein) which doctors use to detect inflammatory processes in our bodies. The user only has to add the reagents and the sample (Figure 7.7)—the device starts automatically, delivering the reagents in sequence to a paper strip where the user visually "reads" the result of the chemical reaction as a colored line, like in a pregnancy test.[52] Another device similarly detects bacteria.[54]

Although not everyone can design these complex circuits, they are as easy to use as your COVID-19 test. Anybody can fabricate them with an inexpensive 3D printer[55] after downloading the designs from a website. The technology is a step forward toward the democratization of medicine. It seems fitting that the same type of trees that, in nearby New England, instigated Benjamin Rush to try to improve the lives of fellow human beings would, from David Juncker's backyard in Quebec, inspire the creation of biomedical devices for all. By learning how plants shuttle fluids using capillary action, we can take advantage of the many juices they have in store for us and further our progress in health care.

Summary

- Trees use capillary action to pump sap up from the roots to the leaves. Small pores in the leaves called *stomata* contribute to the pumping by evaporation, a process called *transpiration*.
- The many natural tubes in a plant that contain sap are called the *xylem*, which can range from a fiftieth of a millimeter in diameter for short plants to a fifth of a millimeter in diameter for very tall trees.
- The *phloem* is a set of vascular structures that runs parallel to the xylem and delivers to the rest of the plant the products generated in the leaves during a cascade of light-triggered chemical reactions (called *photosynthesis*).
- Native Americans had learned to extract sugar from maple trees by "tapping" or making an incision in the bark.
- Similarly, the Olmecs learned to extract latex by tapping the native Panama rubber trees.
- The amount of sugary sap or latex extracted from each tree every day was necessarily small, and both processes were very slow.
- Latex, part of the plant's defense system, runs inside plants through capillary action in small conduits that are separate from the sap-carrying ones.

---

[54] Olanrewaju, A. O., Ng, A., Decorwin-Martin, P., Robillard, A., & Juncker, D. Microfluidic capillaric circuit for rapid and facile bacteria detection. *Anal. Chem.* 89: 6846–6853 (2017).
[55] Karamzadeh, V., Sohrabi-Kashani, A., Shen, M.. & Juncker, D. Digital manufacturing of functional ready-to-use microfluidic systems. *Adv. Mater.* 35: e2303867 (2023).

- Various civilizations that came after the Olmecs—including ours—benefitted from their knowledge on making rubber, which included mixing latex with morning glory vine juices.
- In the Middle Ages, when the Spanish conquistadores conquered the Aztec capital of Tenochtitlán in present-day Mexico, they noticed that the Aztecs knew about latex as well as the medicinal juices of many plants.
- Despite the small amount of latex extracted from each tree, the rubber industry set up large plantations in the nineteenth and early twentieth centuries which enabled the extraction of enormous amounts of latex and had a huge impact on Western society.
- Often inspired by capillarity phenomena observed in Nature, many microfluidic devices are powered by capillary action.
- "Capillaric" devices are user-friendly microfluidic devices that can execute complex assays consisting of preprogrammed steps. Importantly, they can be 3D printed, contributing in the near future to a more democratic access to medicine.

8

Trickling Sand

The Microfluidics of Acequias, Aquifers, and Paleolithic Caves, Next to Microfluidic Models of Sand for Worms

Francisco Vílchez Álvarez is a senior resident from Cáñar, and Antonio Jesús Rodríguez García is an old farmer from nearby Pitres, two neighboring villages in the Alpujarras mountains near Granada, in southern Spain's Andalusia. They have both joined a group of forty or so people who, armed with shovels and pickaxes, are restoring networks of irrigation canals called *acequias* (pronounced *ah-seh-kee-as*)—from the Arabic "as-saqiya," meaning "water conduit"—which were built throughout the area by the Moors centuries ago. "It's a matter of life and death for us," says Antonio. "Without this water, the farmers can't grow anything, the villages can't survive."[1] These acequias are not simple aqueducts—the secret genius and success of their designs lie in the microfluidic properties of the land underneath.

Francisco and Antonio have much in common, as do their small towns. Cáñar lost almost 60% of its inhabitants in the past sixty years and has a population of just 400. Due to the lack of sports fields, a major sports tournament in the village is played against the church's wall. Pitres is even smaller, at 200 residents. In less than thirty minutes, a visitor can stroll across these towns and become enamored with the quaint, narrow streets lined with white-plastered houses, the alleyways covered with wooden beams, and the abundant geraniums lining the patios and windowsills—with no idea of the looming threat that extended droughts pose to this region. Many towns in southern Spain have been hard hit by the rapid increase in temperature brought about by climate change, coupled with the historic practice of intensive farming throughout the region.

The ancient acequias that Francisco and Antonio are re-digging bring much-needed water—and with it, life—to these people. The irony is that there has always been plenty of water in the form of perennial snow in the adjoining Sierra Nevada. Still, without proper irrigation, the land becomes arid. These

---

[1] Méheut, C. Facing a future of drought, Spain turns to medieval solutions and "ancient Wisdom." *The New York Times.* https://www.nytimes.com/2023/07/19/world/europe/spain-drought-acequias.html (2023, July 19).

acequia restoration projects are revitalizing communities decimated by decades of drought in southern Spain and elsewhere.[a] "Now we can grow cherries and kiwis again," says Francisco. Looking down at the green, newly irrigated land, another participant in the restoration project said, "None of this would exist without the acequias. There would be no water to drink, no fountains, no crops. It would almost be a desert."[1]

Through their deep knowledge of mathematics, the Moors became great hydraulic engineers wherever they went. They built waterwheels, dams, cisterns, pools, fountains, aqueducts, and acequias. They came from parched lands and had learned to appreciate every drop of water. Unsurprisingly, in the Quran, Allah proclaims, "We made every living thing out of water." In the Iberian Peninsula, the Moors established the Muslim kingdom of Al-Andalus (eighth century to fifteenth century AD). For more than 800 years, they cultivated an empire that left an everlasting imprint in agriculture, toponymy, and architecture (Figure 8.1).[b] Before the Christians chased them out of the Iberian Peninsula in 1492, the Moors had built approximately 25,000 kilometers (more than 15,000 miles, about one third of all U.S. freeways added together) of acequias across Al-Andalus to irrigate the otherwise dry lands of the hillsides and converted them into terraced fruit gardens.

Farmers used acequias until the 1960s, when globalized agriculture turned to large reservoirs. Without the acequias, the creeks carry the snowmelt down from the nearby Sierra Nevada peaks into rivers and reservoirs that dry up during the summer through evaporation, resulting in a significant loss of utilizable water. The acequias, on the other hand, divert the water gushing down these creeks into multiple channels, spreading it over a larger area.

But acequias do not just *transport* water. Their bottom surface is dirt, so they also *leak* water into the ground by capillarity—they are *designed* that way (Figure 8.2). By seeping water into the soil, acequias "recharge" the earth like a sponge so the land can give water back to the plants and the wells can keep running. Under our feet, vast layers of porous rock and soil completely soaked with groundwater—called *aquifers*—are used by plants (recall from Chapter 7 that an acre of corn can "drink" more than 500 liters *an hour*) as well as by people

---

[a] There is a similar project in New Mexico, where acequias were also once a commonplace irrigation method imported by the Spanish conquistadores. There are close to 700 functioning acequias in New Mexico, and another 20 or so in Colorado.[2]

[b] The name of Spain's southern region Andalusia is derived from the Moorish name Al-Andalus. The Moors built ornate and lavish palaces like the Alhambra and next-door Generalife in Granada as well as the Mezquita in Córdoba, complete with indoor cooling systems, fountains, bath houses, and irrigated courtyards. The indoor Patio de Los Leones in the Alhambra has a system of fountains and water channels designed to cool down the rooms. The Patio de los Naranjos in the Mezquita has a mosaic brick walk that forms a grid of irrigation channels for the orange trees. The courtyard of the Mezquita was built around 976, making it one of the oldest gardens still in existence.

[2] Neuwirth, R. Centuries-old irrigation system shows how to manage scarce water. National Geographic Society. https://www.nationalgeographic.com/environment/article/acequias (2019).

Figure 8.1 Palaces built by the Moors in Andalusia featured delicate architectural motifs combined with sophisticated hydraulic systems. (Top) Patio de la Acequia (Generalife Palace) in Granada, Spain. (Bottom) Patio de los Leones (Alhambra Palace) in Granada, Spain. A marble fountain depicting twelve mythical lions stands at the center of a serene courtyard surrounded by ornately carved columns. The lions' mouths feed four channels in the patio's marble floor—representing the four rivers of paradise—that run throughout the Alhambra palace to cool the rooms down.

Image sources: (Top) User Walter Schärer on Wikipedia. https://commons.wikimedia.org/wiki/ File:Alhambra_-_panoramio_(14).jpg (CC BY-SA 3.0). (Bottom) User Sean Adams on Wikipedia. https://en.wikipedia.org/wiki/File:Palacios_Nazar%C3%ADes_in_the_Alhambra_(Granada). _(51592334991).jpg (CC BY 2.0).

Figure 8.2 Acequia in El Santuario de Chimayo, New Mexico.

Image source: User Dicklyon on Wikipedia. https://en.wikipedia.org/wiki/ File:Potrero_Ditch_at_Santuario_de_Chimayo.jpg (CC BY-SA 4.0).

when they pump water from wells. Approximately one-third to one-half of the flow in an acequia goes to replenish the local aquifer rather than making it to the irrigation field, the river, or the lake. The acequias do not serve just agriculture; they serve the whole community. That's why older people such as Francisco and Antonio, who remember where the ancient ditches were laid out and can help direct their current mapping and excavation, are so vital to acequia restoration projects.

You can experiment with the principles of aquifers yourself. Pour water on dry, sandy soil, and you will notice that the water immediately disappears into the sand until only a wet spot remains. It almost seems to flow down, although the correct term when water quickly finds its way between narrow pores pushed by gravity is *percolation*.[c] (As we will see, clay soil barely percolates, causing different phenomena.) On its way down, water also spreads

---

[c] You probably enjoy the microfluidic marvel of percolation frequently, as it has been exploited by humans for thousands of years. Pour hot water through ground coffee beans or tea leaves, and each drop will trickle down a separate—microfluidic—path, solubilizing a warm drinkable extract that is relished by billions of people every day.

sideways due to capillarity, but not much because there is little time for that. Here, our metaphorical dance between dancers Wetting and Surface Tension must be brief because dancer Gravity is pushy and insists on getting her way. If, on the other hand, you cover a puddle of water with dry dirt, you will observe that the water will ascend against gravity through the soil and will moisten it by capillarity: Here, dancers Wetting and Surface Tension have many alternative paths in between the sand particles to dance their way up and sideways, leaving dancer Gravity with less chance to push them down.

It would be foolish to think that Cáñar's irrigation problems only happen in a tiny, impoverished corner of southern Spain or other arid regions. Just ask the people of Warren, Minnesota, what happened there in the record-hot summer of 2021. Surrounded by irrigated sugar beet fields, the town's well went dry. An older woman explained that she had to drive her riding lawn mower to a neighbor's house after her well went dry. The town officials decided to physically lower the pump almost 20 meters deeper into the aquifer to satisfy the essential drinking and sanitary needs of its 1,500 residents. Similar stories abound of personal-use wells going dry next to irrigation fields pumping 500 gallons a minute.[3]

Aquifers extend over sizable areas under our feet—in California alone, they underlie 40% of the state's land area.[d,4] These aquifers do not look like huge underground lakes—instead, they are more like gravelly sponges—but contain about 30% of Earth's freshwater. By comparison, all the lakes and rivers amount to less than 1%.[e]

Greedy industries are pumping out water faster than natural rain and microfluidic percolation can replenish it.[5,6] About 1.7 billion people worldwide rely on aquifers that are rapidly being depleted.[7] The most depleted aquifer

---

[d] The largest aquifer in the world is located under the soil of Australia and extends over an area that is more than two and a half times larger than France. With an extension of approximately 22% of the country, it provides the only source of fresh water for much of inland Australia. The best-studied aquifers are in the United States, thanks to a *New York Times* study that gathered data from more than 84,000 wells and dozens of agencies to collect millions of groundwater-level measurements, some over a century.[6]

[e] The rest (~ 69%) of the Earth's freshwater is frozen in glaciers and ice caps.

[3] Searcey, D., & Rojanasakul, M. Big farms and flawless fries are gulping water in the land of 10,000 lakes. *The New York Times.* https://www.nytimes.com/interactive/2023/09/03/climate/minnesota-drought-potatoes.html (2023, September 3).

[4] Water Education Foundation. Aquifers. https://www.watereducation.org/aquapedia/aquifers (2023).

[5] Gleeson, T., Wada, Y., Bierkens, M. F. P., & Van Beek, L. P. H. Water balance of global aquifers revealed by groundwater footprint. *Nature* 488: 197–200 (2012).

[6] Flavelle, C., & Rojanasakul, M. Five takeaways from our investigation into America's groundwater crisis. *The New York Times.* https://www.nytimes.com/2023/08/29/climate/groundwater-aquifer-overuse-investigation-takeaways.html (2023, August 29).

[7] Plumer, B. Where the world's running out of water, in one map. *The Washington Post.* https://www.washingtonpost.com/news/wonk/wp/2012/08/10/where-the-worlds-running-out-of-water-in-one-map (2012, August 10).

is perhaps that of the Upper Ganges, which has been heavily overpumped to irrigate farms in northern India and Pakistan. The High Plains aquifer, one of the world's largest, underlies portions of eight states in the central United States—South Dakota, Wyoming, Nebraska, Colorado, Kansas, Oklahoma, New Mexico, and Texas—approximately 174,000 square miles east of the Rocky Mountains in the southern part of the Great Plains. It is the shallowest and most abundant source of groundwater in the region. It supplies drinking water to 82% of the several million people that live on it—and to all the plants that root *in* it. The problem is that the region is also home to around 27% of the irrigated land in the United States and 42% of the nation's fed beef. To irrigate

Figure 8.3 NASA satellite image of circular crop fields in Kansas characteristic of the pivot irrigation technique. The fields shown here are either 800 or 1,600 meters in diameter.

Image source: Wikipedia. https://en.wikipedia.org/wiki/File:Crops_Kansas_AST_20010624.jpg. Public domain.

large areas of land, farmers connect a well to a pump, and the pump to a center pivot—a remotely-operated horizontal mechanical arm on wheels that has calibrated sprinklers and that can be more than half a kilometer long, generating the characteristic crop circles (Figure 8.3). To satisfy consumers' ever-increasing needs—me included—each farm has been pumping out thousands of gallons of water *every minute* for more than sixty years, and there are thousands of wells. In some areas, the aquifer's water level has decreased at a rate of 60 centimeters every year.[7] Many of the wells are dry now.[8]

The residents of Warren, Minnesota, know all too well that they are experiencing a disaster in the making. The Mount Simon–Hinckley aquifer, covering an extensive area of south-central Minnesota and reaching depths of more than 300 meters in some places, contains water that is 30,000 years old. With such an impressive resource, one would assume Minnesota to be a geohydraulic heaven. Yet it is also home to some of the most voracious agribusinesses in the world that have begun to turn it into a geohydraulic hell. Officials had issued more than 7,000 irrigation well permits in Minnesota as of 2022. During the drought of the summer of 2021, farmers cranked their wells up to be able to meet their crops' demands. The law protects the aquifers, but the farmers collectively exceeded what is allowable by permit by more than 6 billion gallons.[3] A third of this overuse happened on the grounds of Offutt Farms, a potato farmer that is a major supplier of McDonald's french fries. Cattle, corn, or soybean farms and textile industries are also very water-intensive, and living near one of these industrial-use wells means that a family's well can go dry.[f] Ironically, Minnesota is known as the "land of 10,000 lakes." Because the bedrock and soil are porous, these beautiful lands are microfluidically interconnected underground. Indeed, the depletion of aquifers is a general trend in almost every state. At this rate, America's status as an agricultural superpower—it is still one of the world's largest exporters of corn, soybean, sorghum, and cotton—will not last long. It is not a hypothetical projection: affected by dried wells, the corn yield has started to decline.[6]

And what's worse, as the wells dry out, refilling the aquifers would require hundreds, if not thousands, of years of rain.[8] Most of the rainfall gets washed to sea without making it to aquifers—up to 40% in the best cases and as little as 8% of rainfall for deeper aquifers.[9] A depleted aquifer can also collapse

---

[f] Fishermen also worry that the trout population is dwindling as the rivers are warming and shrinking as irrigation wells pump out cooler groundwater that usually feeds rivers.[3]

[8] Wines, M. Wells dry, fertile plains turn to dust. *The New York Times*. https://www.nytimes.com/2013/05/20/us/high-plains-aquifer-dwindles-hurting-farmers.html (2013, May 20).

[9] Kotchoni, D. O. V., Vouillamoz, J. M., Lawson, F. M. A., Adjomayi, P., Boukari, M., & Taylor, R. G. Relationships between rainfall and groundwater recharge in seasonally humid Benin: A comparative analysis of long-term hydrographs in sedimentary and crystalline aquifers. *Hydrogeol. J.* 27: 447–457 (2019).

under the weight of the rocks above, shutting the reservoir forever and affecting property value dramatically—in the extreme, it can cause underground collapse or "subsidence" (as already observed in the San Joaquin Valley, California, where water from the wells is now unsafe to drink).[10] Once again: Aquifers are not a giant covered lake that you can easily refill—they are a stack of capillary-interconnected water pores in a mass of compacted dirt and gravel through which it takes centuries for water to percolate from top to bottom. An ultra-fast groundwater flow rate is measured in meters per year (for sandstone and limestone) but can be 1 billion times slower (for compact rocks).[11] Picture your small mound of ground coffee beans, drops of water slowly percolating through it as you patiently wait for your first cup of the morning, and now try to imagine that process occurring literally with a whole mountain of heavily compacted, ground coffee beans—you would have to wait too many lifetimes even to see that first drop make its way through to the bottom of the proverbial cup.

In summary, once we have depleted the aquifers, they will cease to exist for us. We are standing on fragile capillary beds. Our planet will no longer be able to supply enough water both for us to drink and for environmentally hostile industries to operate. The microfluidic nature of our soil dictates that our only chance of survival is by urgently demanding that all our industrial practices be sustainable on all levels.

You see that capillarity is essential to preserving water in areas where it is a precious resource. This principle extends beyond the storage of water in aquifers. As soon as a water drop contacts the surface of dry land due to morning dew or rainfall, it is quickly wicked under the surface layers. Under the sun, the dry surface soil acts as an insulation against the blasting thermal energy of the sun's rays, protecting the underground water from evaporating in the day's scorching heat. The surface looks barren to us, but a community of burrowing organisms such as beetles, snails, cockroaches, spiders, and scorpions inhabits the underground of dry zones. These creatures obtain the little water they need from the moisture and food they find under the surface. It's almost as if our planet knew that water is unique for safeguarding its ecosystems and had devised a sponge-like mechanism to preserve every drop below its surface. And we humans—who understand microfluidics and presume to be the most intelligent species on

---

[10] James, Ian. California cracks down on another Central Valley farm area for groundwater depletion. Los Angeles Times. https://www.latimes.com/environment/story/2024-09-21/california-puts-another-farm-area-on-groundwater-probation (2024, Sept 21)

[11] Ferguson, G., & McIntosh, J. C. Groundwater—not ice sheets—is the largest source of water on land and most of it is ancient. The Conversation. https://theconversation.com/groundwater-not-ice-sheets-is-the-largest-source-of-water-on-land-and-most-of-it-is-ancient-174031 (2022, January 19).

Earth—are, paradoxically, pumping out water from the sponge we live on at a rate that threatens our civilization.

Proper water management to save water is needed not only in times of drought but also every time it rains. In many areas, mainly when asphalt or cement covers large surfaces, such as in cities and suburbs, the opposite phenomenon—termed *surface runoff*—happens: Excess rainwater, stormwater, meltwater, or other sources cannot sufficiently infiltrate the soil. Flooding occurs, sometimes with devastating consequences. If you live near a lake, you are familiar with the curbside drain sign that reads "Drains to lake," a warning that the unfiltered stormwater running down the street can have negative environmental consequences. To avoid surface runoff, urban developers in the United States and the United Kingdom build many new housing and parking lots with a shallow ditch of dirt or plants around them called a *swale*. A swale acts as a capillary sponge that absorbs and filters rainwater—as a dead-end acequia, as it were—only during heavy rainfalls. Swales help prevent floodings and, like their acequia cousins, help replenish groundwater, restoring natural aquifers. They are smarter and more aesthetically pleasing than cement parking strips. Hopefully, in the not-so-distant future, these microfluidically powered landscaping features will become commonplace throughout our cities.

You may be wondering why rivers and lakes do not slowly empty through their beds because they are made of soil, too. The answer resides in the microfluidic properties of the soil below. Soil is a layer of mineral particles formed by the weathering of rocks and itself lying over the bedrock. It can contain gravel, rocks, sand, silt, clay, loam (a mixture of sand, silt, and clay—the "gardener's soil"), and organic matter (humus). Sand, silt, and clay differ in their ranges of particle sizes: 50 micrometers to 2 millimeters for sand, 2–50 micrometers for silt, and less than 2 micrometers for clay. Indeed, the soil at the bottom of lakes and rivers is heavily soaked with water, and deep below, there is a layer of impermeable rock, but clay plays a significant role here, too. Clay may seem like a very porous powder when dry. However, it is formed of tiny plate-shaped particles that, in the presence of water and with the help of high pressures at the bottom of a lake, align and adhere very tightly to each other, closing all the pores and stopping the water from flowing deeper. Each platy microparticle of clay effectively functions like a microfluidic check valve—a microscopic version of the flap valve you see when you open your toilet's reservoir.

The extremely low hydraulic permeability of clay has impacted humanity in ways that go far beyond aquifers and agriculture. We see an example when we walk into the Neanderthal caves at Lascaux, a small locality in the Dordogne region near Montignac, on the banks of the Vézère River in southwestern France.

Percolating water has eroded this 100-million-year-old limestone hillside into a landscape of underground galleries called *karst cavities*.[g] Anatomically modern humans (*Homo sapiens*) had been roaming around Europe since at least 40,000 BC. About 18,000 years ago, early dwellers discovered a large cave in Lascaux formed by underwater currents and took refuge in one or more of these cavities from the inclement weather of the late glacial period.

But these karst cavities were special. The caves were protected from further erosion by rainfall by clay sediments and rocks *on top* of the caves. "One reason [the cave] was protected is there is a layer of clay in the soil that waterproofs the cave. That's why Lascaux has no stalactites or stalagmites. It's a dry cave," explained Guillaume Colombo, the director of the new cave and museum complex at Lascaux.[12] This microfluidic interplay between sand and water produced a unique shelter for our European ancestors that lasted for many generations. Protected by this impermeable geological roof, cave dwellers had time to paint (Figure 8.4, top),[h] cook, chat under the flickering light of a bonfire or a rudimentary oil lamp, and perhaps sing some of the first human songs.

The primitive inhabitants of the late Paleolithic period who used stone tools noticed the unique properties of clay and used it to make pottery at least 30,000 years ago. During the Neolithic era, their descendants made clay bricks to build the first settlements. These ancient toolmakers must have also learned that a small part of the water in a dried clay bottle was lost to its walls because its exterior felt humid. They noticed the water stayed cold even on the hottest days: Evaporation at the wall surface lowered the container's temperature. Many drinking devices still use this principle (Figure 8.4, bottom). These human artifacts are slightly permeable because not having been exposed to high pressures, the microparticles of clay are not perfectly aligned—that is, the "check valves" are not fully closed. When humans built their first food and drinking containers, houses, and towns, they unknowingly relied on microfluidic interactions between tiny grains of sand.

Throughout millennia, rivers have carved valleys out and deposited deltas made of moist, fertile soil from their sediments. Subsurface layers of clay—deposited by the rivers eons ago—stop the water from percolating down from

---

[g] A famous example is the Škocjan Caves in Slovenia, with spectacular stalactites and stalagmites resulting from the continuous percolation of water through the limestone over the ages.

[h] Due to the dryness of the walls, they were able to use magnesium pencils and blow-drying of ochre pigments with hollow bird bones.[12] Earlier rock art paintings that depict animal forms and handprints have been discovered elsewhere, most notably in the caves of Altamira, northern Spain (~35,000-year-old paintings), and in the Maros-Pangkep caves on the Indonesian island of Sulawesi (~40,000- to 45,000-year-old paintings).

[12] Beardsley, E. Next to the original, France replicates prehistoric cave paintings. NPR. https://www.npr.org/sections/parallels/2017/01/02/507549682/next-to-the-original-france-replicates-prehistoric-cave-paintings (2017, January 2).

Figure 8.4 (Top) Painting at the Lascaux caves. (Bottom) A water container made with clay that cools down its contents by evaporative cooling and capillary wicking through the walls.

Image source: Wikipedia user Oricalve (public domain). Image sources: (Top) Wikipedia user Jack Versloot. https://en.wikipedia.org/wiki/Lascaux#/media/File:Lascaux_II.jpg (CC BY 2.0). (Bottom) Wikipedia user Oricalve. https://ca.wikipedia.org/wiki/C%C3%A0ntir#/media/Fitxer:Cantir_bisbal.JPG. Public domain.

the surface when it rains, and capillary action recirculates it laterally and upwards through the more porous soils. (Occasionally, underground creeks or currents also form.) These agents play a critical role in cooperating to fertilize riverbanks by retaining and spreading the rainwater and seeds. On these rich mineral deposits of riverbanks, humans learned to raise their first crops, domesticated animals, and settled on incipient villages during the Neolithic era, abandoning their precarious hunter–gatherer existence in caves. An example is the ancestors of the Olmecs in Mesoamerica, who understood that the soil of the basins of the Papaloapan, the Coatzacoalcos, and the Tonalá was ideal for growing corn and tapping latex. There they learned to exchange their goods and crafts. Similarly, the region known as the Fertile Crescent, fed also by three rivers—the Euphrates, the Tigris, and the Nile—became a land of rich agriculture and trade around 9,000 BC, then the cradle of the Sumer Empire in Mesopotamia starting in 4,500 BC and many other cultures afterward. In addition to agriculture, it is here that writing, the wheel, irrigation, and glass were invented.

* * *

A vital and surprising reason soil is so porous to fluids is that an unfathomable quantity of crawling organisms inhabit it—including the roots of plants, fungi, ants, worms, millipedes, beetles, bumblebees, and more. Everything gets composted underground. When these life forms die, they are rapidly eaten by others and are processed by bacteria.[i] They leave behind precious, empty, dark spaces, tiny pores in which other life forms will cohabitate, reproduce, and die. Water will periodically flood them each time rain wicks through the ground. In the dunes of arid areas, scientists have observed air and water vapor penetrating through the pores of large volumes of sand as if the dunes were purposely inhaling and exhaling. This air movement condenses the little water floating in the air and helps sustain microscopic life in the sand.[13] Dig in anywhere around you, and you'll conclude that Earth is like a sponge, alive with critters.

Among these living creatures, one is startlingly abundant: the roundworms, which scientists call *nematodes*, constantly digging around. It's their home.

---

[i] Bacteria grow by billions in every gram of soil and have been found at subterranean depths of more than 5,000 meters.[14]

[13] Louge, M. Y., Valance, A., Xu, J., Ould El-moctar, A., & Chasle, P. Water vapor transport across an arid sand surface—Non-linear thermal coupling, wind-driven pore advection, subsurface waves, and exchange with the atmospheric boundary layer. *J. Geophys. Res. Earth Surf.* 127: e2021JF006490 (2022).

[14] Watts, J. Scientists identify vast underground ecosystem containing billions of microorganisms. *The Guardian.* https://www.theguardian.com/science/2018/dec/10/tread-softly-because-you-tread-on-23bn-tonnes-of-micro-organisms (2018, December 10).

Not to be confused with the larger, slimy earthworms, most nematodes are 1 or 2 millimeters long—they are barely visible. Each nematode can lay thousands of eggs. Hence, if you dig a cubic meter of soil and look under the microscope for the tiny ones, researchers estimate you will find more than 1 million individuals.[j] Believed to be one of the oldest organisms on Earth, they have adapted to live in the Arctic Circle; on the ocean floor (where they constitute 90% of the animals[15]); in salty arsenic-poisoned lakes; and as parasites inside plants and animals, including humans.[17] Nematodes are the most abundant animal on Earth—four of five animals on the planet are nematodes. There are an estimated 400,000,000,000,000,000,000 (that's *four hundred billion times one billion*) nematodes on Earth's topsoil, so there are roughly 50 billion soil nematodes for every human inhabiting our planet.[18] They have even been found in tap water.[19] It's their world; you just happen to stand on top of it.

Cornelia Bargmann, who goes by "Cori," is a star neuroscientist at the Rockefeller Institute in New York who developed an early fascination for these nematodes and how they interact with their soil environment. She credits her mother for piquing her interest in the behavior of animals. "My mother had books by Konrad Lorenz[k] and other early neuroethologists that she would read to us. I still have her books," she told me. She became a world expert in the nematode *Caenorhabditis elegans*, which usually feeds on bacteria and lives in dirt and rotting fruit.

These worms may not seem very interesting initially, but they have several advantages, making them the rockstar-like model organism for thousands of biologists. For starters, they are inexpensive to maintain and feed. They are transparent, so scientists can observe their cells under a microscope as they grow and divide. Their optical clarity has allowed scientists to painstakingly

---

[j] The nematodes can be a pest to some human-grown plants such as the soybean, so people count the number of eggs in the soil to determine the level of infestation. A soil sample of 100 cubic centimeters is deemed to have high infestation if it contains 5,000 eggs.[16] That would be equivalent to 50 million eggs per cubic meter of soil.

[k] Konrad Lorenz (1903–1989) is considered one of the founding fathers of modern ethology, the study of animal behavior. In 1973, he was awarded the Nobel Prize in Physiology or Medicine for his ethology studies.

[15] Danovaro, R., Gambi, C., Dell'Anno, A., Corinaldesi, C., Fraschetti, S., Vanreusel, A., et al. Exponential decline of deep-sea ecosystem functioning linked to benthic biodiversity loss. *Curr. Biol.* 18: 1–8 (2008).

[16] Kelly, H. M. Time to sample soil for pathogenic nematodes. *UTcrops News Blog*. https://news.utcrops.com/2021/11/time-to-sample-soil-for-pathogenic-nematodes-2 (2021, November 9).

[17] SciShow. The most important animal you've never seen: Meet the nematode. YouTube. https://www.youtube.com/watch?v=vBWzrlCBhCM (2020).

[18] van den Hoogen, J., Geisen, S., Routh, D., Ferris, H., Traunspurger, W., Wardle, D., et al. Soil nematode abundance and functional group composition at a global scale. *Nature* 572: 194–198 (2019).

[19] Buse, H. Y., Lu, J., Struewing, I. T., & Ashbolt, N. J. Eukaryotic diversity in premise drinking water using 18S rDNA sequencing: Implications for health risks. *Environ. Sci. Pollut. Res. Int.* 20: 6351–6366 (2013).

"nickname" each cell of the organism with a three-letter code. The animal has exactly 302 neurons, which doesn't seem much by the standards of a human brain. Still, those are enough to "learn" certain behaviors that the researchers can then manipulate by changing specific three-letter-coded cells with lasers or genetic techniques. Scientists working with mice, for example, do not have the requisite understanding of the—much more complex—mouse neuronal networks to perform this type of cellular-level manipulation. Also, *C. elegans* only live for two weeks, so you do not need to wait long to see experimental results. It's like having a worm made of mini-LEGO pieces and testing whether it wiggles differently—or lives longer—after swapping out any part you want.

One day in 2004, Cori Bargmann realized that microfluidic devices and dirt had many things in common. Hang Lu, a Chinese postdoctoral student in her lab,[1] asked whether it would be possible to introduce live worms inside a microfluidic device made of a transparent rubber and examine their behavior.[20,21] Could the rubber channels effectively simulate the pores in the soil through which worms usually crawl in their natural habitat? If so, the worms could be observed and manipulated under a microscope with a new level of precision. Because vision is not helpful in the darkness of their underground world, worms have not evolved to have eyes like us. Instead, they have a keen sense of olfaction (smell) that helps them navigate their unilluminated but richly scented world. In 2007, Nikos Chronis, a Greek postdoc,[m] also joined the team.

Bargmann studied worms for years to elucidate the many molecular wonders and mysteries of learning and olfaction that worms share with us. One problem was that it had been difficult to experimentally manipulate the underground environment of nematodes. Bargmann had the intuition that she needed a microfluidic approach—and engineers to apply it. The microfluidic devices that Lu and Chronis prepared were transparent, made of soft rubber material, and could be accurately designed using techniques borrowed from the microelectronics industry.

More than 100 labs worldwide have followed this simple yet powerful paradigm started by Bargmann and her engineers. Much like photographers used to develop film, Lu and Chronis made the negative mold of the channels in a light-sensitive polymer, a process termed *photolithography* (see Figure 2.4). Then, they replicated the mold in transparent rubber (see Figure 2.5). Lu used the devices to deliver stimuli and observe free-moving behavior. Chronis

---

[1] Hang Lu is now a professor at the Georgia Institute of Technology.

[m] Nikos Chronis is now a professor at the National Technical University of Athens.

[20] Gray, J. M., Karow, D. S., Lu, H., Chang, A. J., Chang, J. S., Ellis, R. E., et al. Oxygen sensation and social feeding mediated by a *C. elegans* guanylate cyclase homologue. *Nature* 430: 317–322 (2004).

[21] Zhang, Y., Lu, H., & Bargmann, C. I. Pathogenic bacteria induce aversive olfactory learning in *Caenorhabditis elegans*. *Nature* 438: 179–184 (2005).

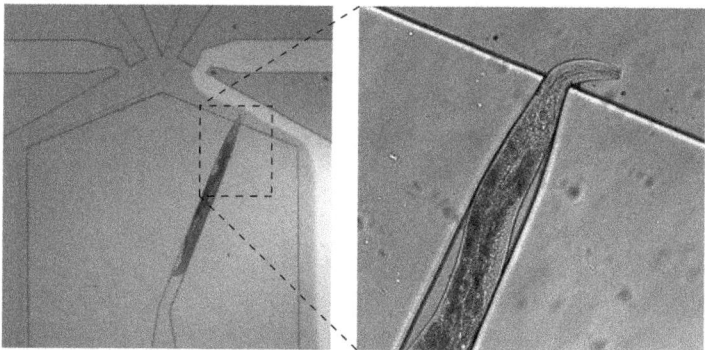

Figure 8.5 A nematode worm (dark) trapped inside a microfluidic channel and extending its head into a second channel where the researcher can fluidically modulate the composition (white fluid). At its widest, the worm is approximately one-tenth of a millimeter (100 micrometers), and the opening at the end of the microchannel is about 40 micrometers wide.

Image courtesy of Nikos Chronis, National Technical University of Athens.

precisely designed a microfluidic channel so that the worm would crawl through one end and get stuck at the other, sticking just the tip of its head—which acts as its "nose"—into another channel (Figure 8.5). In this second channel, Chronis switched flows on and off with a valve, resulting in the worm twisting its nose left and right to smell the chemical stimuli. At the same time, the transparent material allowed Chronis to observe signals telling him which nerve cells in the worm's brain were activated by the compound. Notably, the device could be reused and worked the same way for all worms because they all reached the exact same size in adulthood. Bargmann has received numerous awards for her work, including the Kavli Prize and the Breakthrough Prize in Life Sciences. "Many years later, I asked Hang how I had the good luck that she joined the lab," Bargmann said. "Hang gave me a wry look and said she had written to 50 biologists, and I was the only one who answered."

This tool led to similar ones, and rubber microfluidics became a new standard for biologists to mimic the natural habitat of tiny organisms under the microscope. These transparent microfluidic devices designed to capture and trick worms also served, as envisioned by Bargmann, to free worm biologists from the veil of dirt. The approach has been applied to other organisms and cells, from bacteria to fly embryos and sperm.

The tool befits its function. Throughout the ages, microfluidics has played a role in percolating water stories that predate the depletion of aquifers in Andalusia or Minnesota—from nomads settling around fertile, clay-rich riverbeds to modern-day agriculture and science. Like worms and plants, our civilization is rooted in a capillary bed.

Summary

- More than 1,000 years ago, to help distribute snow runoff or river water to distant fields for irrigation in what is now southern Spain, the Moors dug canals called *acequias*. Acequias, which have a system of gates to regulate flow and require the attendance of maintenance personnel, are built and maintained in many other areas of the world.
- Because the bottom of an acequia is made of permeable dirt, the water it transports also seeps into the ground by capillarity and helps maintain the groundwater levels of aquifers.
- Humans have extracted water from aquifers through wells for millennia, and until now the capillary-driven process of replenishment with rainwater was sufficient. Recently, aquifers in many areas of the world (including the United States) are being overpumped to feed agricultural irrigation and industries at rates exceeding the natural rainwater replenishment rates to the point where people cannot get drinkable water anymore.
- In urban planning, to prevent floodings in areas with a lot of cement or asphalt, a swale—a shallow channel with sloping sides—is built on the side of the road or a parking lot to manage water runoff and increase rainwater infiltration.
- The microfluidic interplay between sand and water over millions of years produced erosions in (porous) limestone layers that are covered with (impermeable) clay deposits, leading to karst cavities such as the Lascaux caves that were inhabited by humans some tens of thousands of years ago.
- The behavior of the worm *C. elegans*, which normally lives and feeds in dirt and is an important model organism in genetics and cell biology, has been observed and manipulated with microfluidic channels that mimic the micropores between grains of dirt.

9

Nadal's Sweat

The Microfluidics of Sweat-Based Cooling, Our Various Glands, and Wearable Sweat Sensors

Rafael Nadal could not have guessed that after playing for nearly five-and-a-half hot-and-humid hours in the 2022 Australia Open, he would be assailed with questions about his profuse sweating—despite being a very famous, charismatic tennis player accustomed to having much more than the average renowned person's share of media and paparazzi attention. It is not unusual to see his shirt clinging to his body (Figure 9.1), shining like latex and drenched in sweat—which, after all, is part of our built-in microfluidic cooling system—but he had been sweating way more *than usual*.

Figure 9.1 Rafael Nadal returning a ball.

Image source: User Mike McCune on Flickr. https://www.flickr.com/photos/51035597937@N01/ 6842480972 (CC BY 2.0).

The game was played in the southern tip of Australia, in Melbourne's Rod Laver Arena, jam-packed with fans thrilled to see the unfailing, energy-filled player hit the never-reached-before mark of twenty-one Grand Slam men's titles. He dripped so much sweat that the floor sweepers had to wipe around him constantly during breaks to prevent the acrylic floor from becoming too slippery. Because of the sweat, he could not keep the tennis balls in his pockets as players usually do. His opponent did not sweat nearly as noticeably. At the end of the game, his doctors advised him to jump on an exercise bike *to cool down* before the press conference. "Is this how superheroes sweat?" the press wanted to know.

There is a bit of truth in that. As it turns out, there is an evolutionary advantage to sweating. But the predominant theory of human evolution—as allegorically staged in the opening act of Stanley Kubrick's science-fiction legendary film *2001: A Space Odyssey*—posited until recently that humans diverged from apes when they learned to handle tools. Although in the study of prehistory it is difficult to discern cause from effect, it kind of made sense. According to this theory, tools conferred humans a hunting advantage, which translated into a survival dominance. Their use forced humans to stand up, freeing some muscle-control areas in the brain for more creative functions such as language. However, this theory has major flaws. The first hominins[a] with clear signs of bipedalism (movement on two feet)—such as *Lucy*, the famous female *Australopithecus afarensis* fossil from between 3 and 4 million years ago in the Awash Valley in what is now Ethiopia—had a tiny brain about one-third of ours. It was not until 1 or 2 million years ago—between the *Homo habilis* and the more recent *Homo erectus*, all expert walkers—that the brains of our ancestors started to grow very fast. And there is a long list of animals that also use tools—the great apes (gorillas, chimpanzees, and orangutans), monkeys, otters, dolphins, octopuses, corvids (which includes crows, ravens, and rooks), and some crocodiles—but their brains have not grown like ours, so tool usage cannot have been the *only cause* of the emergence of superior intelligence.[b]

A newer theory of human evolution connects bipedalism with sweat[c] and leaves brain growth as an independent evolutionary process.[1] This theory starts by noticing a coincidence: Humans are one of the rare non-avian

[a] The term *hominin* designates modern humans (*Homo sapiens*) and all of our extinct bipedal ancestors—those ancestors who walked upright on two feet, such as *Australopithecus* and *Neanderthals*.

[b] Note that brain size and intelligence grew together during the *evolution* of most vertebrate species, but neither brain size nor brain-body mass ratio are correlated with intelligence when comparing *different* species.

[c] Of note, bipedalism might have had many concomitant causes, such as the need to carry objects/food (increasing the efficiency of foraging), to stand tall above the grass for vigilance, or to reduce sun exposure, among others.

[1] Lieberman, D. E. Human locomotion and heat loss: An evolutionary perspective. *Compr. Physiol.* 5: 99–117 (2015).

animals—along with kangaroos and running lizards—that use bipedalism for locomotion and also one of the rare animals that sweat—along with apes, monkeys, horses, and zebras—but humans are the only ones that do both. According to this theory, the combined advantages of bipedalism and sweating allowed humans to hunt animals by running them to exhaustion in the scorching heat of the African savannah, where the first hominins appeared.[1]

This "savannah hypothesis" makes physiological sense: Sweating is a thermoregulatory process of secreting fluids through tiny (microfluidic) vessels that end in pores in our skin to cause evaporative cooling. It would also explain why we have so little fur compared to other primates, as fur impedes evaporation. Every molecule of water that evaporates takes heat into the atmosphere and lowers the temperature of the surface left behind—in our case, our furless skin. Who would have guessed that microfluidics gave a boost to human evolution?

The human body cools down using microfluidic conduits in the form of sweat glands that drain into pores covering all our skin. The palms and soles of our feet have the greatest density of cooling, or "eccrine" sweat glands, approximately 600–700 glands per square centimeter. The glands sit a few millimeters deep under the skin. The secretory ducts that open to the skin can be 4–8 millimeters long and have diameters ranging from 10 to 20 micrometers,[2] ending in pores too small to be visible to the naked eye. These pores line up in a single file between the ridges of fingerprints (Figure 9.2). Other glands, the "apocrine" glands, connect directly to the hair follicles in the armpits and genital region and secrete a thick fluid that produces "body odor" after contact with skin bacteria. The underarm also has a high concentration of eccrine sweat glands, so we sweat abundantly there. The brain controls sweating through neurons that directly innervate each sweat gland. Thermosensitive neurons in the brain and the skin detect the internal body and skin temperatures, so when we enter a hot environment or start physical activity, the brain instructs the sweat glands to release sweat to maintain the body temperature at approximately 37°C. Essentially, the skin is a microfluidic cooling machine that wraps our whole body.

This ultra-thin refrigeration system has other functions, at least for nonhuman mammals. In addition to sweat, an animal's body also exudes minute quantities of volatile chemicals called *pheromones* that can modify the behavior of others toward them—including their aggression and sexual attraction—without their conscious knowledge. Scientists believe that the deep, instant bond between a mother and her infant is mediated, in part, by these chemicals. In humans, the effects of pheromones are still controversial because they can be fogged by complex cultural signals, clothing, and perfume. But they are very compelling in animals, on which we can more easily perform experiments to isolate and dissect the pheromones' effects.

---

[2] Rabost-Garcia, G., Farré-Lladós, J., & Casals-Terré, J. Recent impact of microfluidics on skin models for perspiration simulation. *Membranes* 11: 1–13 (2021).

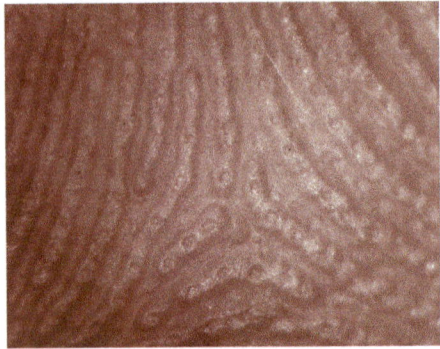

Figure 9.2 Human sweat gland pores on the ridges of a finger pad.

Image source: User Katsamenis on Wikipedia. https://en.wikipedia.org/wiki/ Sweat_gland#/media/File:Fingerprt.jpg (CC BY-SA 3.0).

There are dozens of other glands in the body with diverse functions, all using microfluidic ducts to deliver their load. Among the most critical, *salivary glands* produce between 0.5 and 1.5 liters of saliva daily. Saliva lubricates the food in preparation for swallowing, prevents the internal mouth skin from drying, and contains essential enzymes that help initiate digestion and avert tooth decay. We would not live for very long without these microfluidic marvels. Glands in the stomach and pancreas produce the acid and digesting enzymes necessary to transform the proteins, fats, and carbohydrates of food into nutrients. The *prostate glands* produce semen, the liquid emitted during the male ejaculatory response in mammals, ensuring successful reproduction. *Sebaceous glands* secrete an oily fluid (called *sebum*) through the hair follicle to help lubricate the hair and skin. The production of excess sebum is the cause of acne, a common problem during puberty in teenagers. That pesky earwax comes from the *ceruminous glands* not as a nuisance but to protect the eardrum from water, dust, fungi, and bacteria. Tears flow from one *lacrimal gland* above each eye, keeping the eyes' surface moist—and overflowing when emotion demands it. In all mammals, a mother's *mammary glands* secrete the milk that feeds her offspring. As a parent and microfluidics engineer, I find it reassuring that, at some point in evolution, natural selection "chose" microfluidics as the most reliable way to deliver nutrients to the progeny.

How is sweating regulated in our body? The master sensor and switch are both in the hypothalamus, a tiny organ the size of an almond situated in the brain. The hypothalamus is another microfluidic wonder, and it is responsible for regulating key processes such as hunger, thirst, blood pressure, fatigue, sleep, and sexual behavior, as well as body temperature and other critical functions. A small, specialized network of blood capillaries perfuses parts of the hypothalamus, bringing information about the rest of the body to the brain. Nerve cells or neurons "read" relevant information from these capillaries and release necessary hormones into them to help maintain a balance of body functions. The hypothalamus is like a microfluidic physicochemical analyzer and Nintendo controller. With this small-analyzer strategy, our body can react quickly. When our body gets one or two degrees hotter than ambient temperature, the warm-sensitive neurons of the hypothalamus relay a "too hot" signal to other neurons that connect to the spinal cord—the long-distance information highway of our body for electrical signals—which distributes the signal to neurons next to the sweat glands of the skin. These last neurons spit out *acetylcholine*, a chemical messenger that tells the sweat glands to start pumping sweat through the channels that end in our skin's tiny pores. The information transfer process might seem rather complex and involved, but it can happen in seconds when you enter a sauna or start feeling nervous at a job interview. The response is so fast because the temperature sensor (the hypothalamus) and the cooling units (the sweat glands) are tiny. That's the magic of microfluidics.

Despite not knowing any biology or physics, prehistoric hunters realized millions of years ago that they could outrun a gazelle or an antelope by patiently chasing after them for hours under the sun.[d] Unbeknownst to them, they carried high-tech gear by evolutionary standards: an ultra-light, wearable microfluidic refrigerator (the skin and sweat glands) regulated by a microfluidic biological microcontroller (the hypothalamus). Without these sophisticated advantages, the animal eventually had to stop to cool off and rest. At that point, the heat-shocked creature was at the mercy of microfluidically cooled humans carrying stones, clubs, and spears. These hunters needed to cooperate and make tools, which contributed to their brain growth and made them master hunters of the savannah—no longer prey. In these excursions, they likely ran into other tribes and learned to exchange knowledge and build larger societies. By equipping the human species with cutting-edge microfluidics, evolution propelled it through the grind of natural selection.

* * *

Ultimately, sportspeople such as Nadal would like to wear a thin sensor on their wrist that alerts them about the ups and downs of their metabolism's physiology based on their sweat. The engineer who first came up with the idea of combining electronics with plastics to develop *flexible electronics* was John Rogers, Professor of Materials Science and Engineering at Northwestern University. Making electronics thin and flexible was a deliberate choice to make them *wearable*. "To be able to put electronics on the human body occurred to us as being a richer area for research and a lot more promising for broad societal impact than a new piece of consumer electronic gadgetry," Rogers said.[4]

---

[d] This strategy is still practiced by tribes in the Kalahari of Africa and into the 20th century in other locations.[3]

[3] Liebenberg, L. Persistence hunting by modern hunter–gatherers. *Curr. Anthropol.* 47: 1017–1025 (2006).

[4] Caine, P. Northwestern engineering team pioneers new medical technologies. WTTW. https://news.wttw.com/2019/09/09/northwestern-engineering-team-pioneers-new-medical-technologies (2019, September 9).

Rogers' colossal success has hinged on exploiting a plain microscale effect: A microelectronic chip's active part—where electrons are flowing—consists of an exceedingly thin layer confined to the electronic chip's surface—only a fraction of a micrometer thick. Below the active surface, the inactive silicon body of the chip, which is about a thousand times thicker than the active part, is as electronically dead and silent as a rock. Therefore, as Rogers argued more than fifteen years ago, one might as well get rid of the inactive silicon body and grow, print, or mount the active electronic layer on flexible polymer films to exploit the benefits of plastics. Rogers stated, "We were previously working on thin polymer semiconductors on plastic sheets as the basis for flexible circuits—so maybe it was natural to think about the active stuff on the top surface and the mechanical support below. We just needed to replace the wafer as a mechanical support with our plastic sheets, retaining the thin silicon near the surface."[5] Plastics are transparent, biocompatible, and can be bent or inflated into various shapes. A pediatrician can gently apply a soft temporary "tattoo" with thin-film electronics to the skin of a neonate to monitor their vital signs wirelessly.[6] A heart surgeon can introduce into a patient's arteries balloon catheters with electrodes that provide real-time information on the arteries' health.[7]

Rogers has extended this idea to "wearables" that incorporate thin (<1-millimeter-thick) microfluidics for sampling sweat. An array of integrated microelectronic thin-film sensors measure temperature and heart rate, although to "see" the results, you need to connect your smartphone via Wi-Fi. In 2016, his group demonstrated a temporary tattoo whose thin rubber microfluidic channels, sealed onto the skin's surface, could harvest sweat from the skin's pores.[8] Using a combination of capillarity and the natural pressure generated by perspiration, the device then routes the sweat to various reservoirs. In each reservoir, a chemical reaction changes the color of the solution to detect the multiple ions present in sweat: chloride, hydronium ions, glucose, and lactate. A sweat patch incorporating capillary burst valves allowed twelve timed perspiration samples (one per minute), cleverly organized on the sensor in a watch-like manner (Figure 9.3).[9] This sensor became the basis of Gatorade's Gx Sweat Patch, a single-use wearable microfluidic device that measures your sweat rate and the

[5] Folch, A. *Hidden in Plain Sight: The History, Science, and Engineering of Microfluidic Technology.* MIT Press, 2022, 139.

[6] Chung, H. U., Rwei, A. Y., Hourlier-Fargette, A., Xu, S., Lee, K., Dunne, E. C., et al. Skin-interfaced biosensors for advanced wireless physiological monitoring in neonatal and pediatric intensive-care units. *Nat. Med.* 26: 418–429 (2020).

[7] Kim, D.-H., Lu, N., Ghaffari, R., Kim, Y. S., Lee, S. P., Xu, L., et al. Materials for multifunctional balloon catheters with capabilities in cardiac electrophysiological mapping and ablation therapy. *Nat. Mater.* 10: 316–323 (2011).

[8] Koh, A., Kang, D., Xue, Y., Lee, S., Pielak, R. M., Kim, J., et al. A soft, wearable microfluidic device for the capture, storage, and colorimetric sensing of sweat. *Sci. Transl. Med.* 8: 366ra165 (2016).

[9] Choi, J., Kang, D., Han, S., Kim, S. B., & Rogers, J. A. Thin, soft, skin-mounted microfluidic networks with capillary bursting valves for chrono-sampling of sweat. *Adv. Healthc. Mater.* 6: 1601355 (2017).

concentration of sodium in your sweat. The Gx Sweat Patch was released in 2021 and is manufactured by the millions.

Colorimetric reactions, however, are not very quantitative. Researchers use chemical reactions controlled by electrical currents called electrochemical reactions to obtain more exact measurements.[10] Electrochemical reactions can also be more specific, but they require electrical power. Chemists have started incorporating tiny batteries, which generate electricity from a chemical reaction, into the sensors to help power them. Joe Wang's group at the University of California, San Diego has developed stretchable "biofuel" cells that, printed on fabric, generate power for wearable sensors from a chemical reaction with the perspiration itself: socks that light up when you start running, and so on.[11] For women who need to check their levels of estradiol—a hormone that indicates how well the

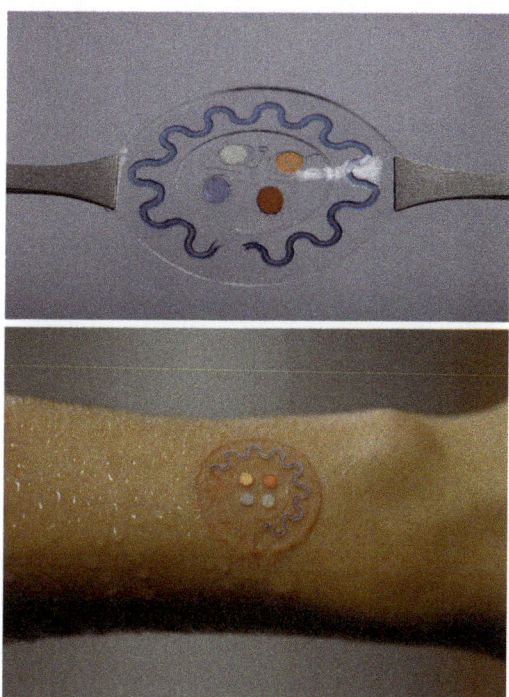

Figure 9.3 Microfluidic wearable for chrono-sampling and analyzing of sweat.
Images courtesy of John Rogers, Northwestern University.

[10] Bandodkar, A. J., Gutruf, P., Choi, J., Lee, K., Sekine, Y., Reeder, J. T., et al. Battery-free, skin-interfaced microfluidic/electronic systems for simultaneous electrochemical, colorimetric, and volumetric analysis of sweat. *Sci. Adv.* 5: eaav3294 (2019).

[11] Jeerapan, I., Sempionatto, J. R., Pavinatto, A., You, J. M., & Wang, J. Stretchable biofuel cells as wearable textile-based self-powered sensors. *J. Mater. Chem. A. Mater. Energy Sustain.* 4: 18342 (2016).

ovaries, placenta, or adrenal glands work and that is used to monitor fertility cycles—Wei Gao's lab at Caltech developed a finger-wearable electrochemical microfluidic sensor that "reads" the concentration of estradiol in sweat from fingers (Figure 9.4). The ring is also chemically powered by sweat and connects to a smartphone app via Wi-Fi.

Incredibly, these devices are as thin as your skin and are powered by nothing but capillary action and the salty fluid that natural selection chose to cool us off.

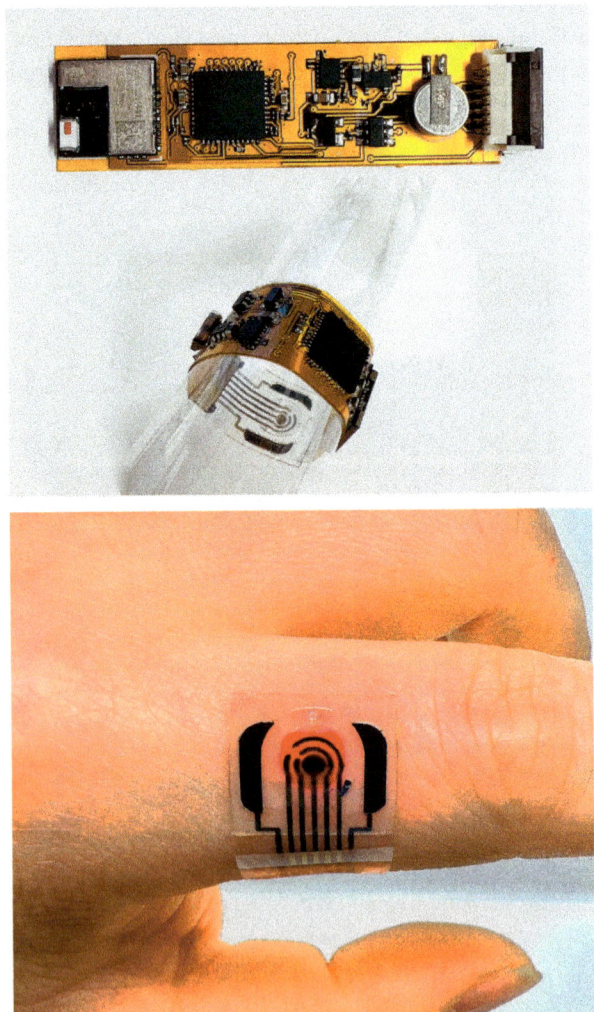

Figure 9.4 Finger-wearable microfluidic sweat sensor to monitor the levels of the female fertility hormone estradiol.

Images courtesy of Wei Gao, Caltech.

Watching Nadal grunt and sweat, one cannot stop wondering whether he might be a fortuitous mutant pushing our species up the ladder of human evolution.

Summary

- Our body has dozens of secretory *glands* with diverse functions, all using microfluidic ducts to deliver their load.
- Distributed all over our skin, sweat glands are our most abundant type of glands and are a microfluidic cooling mechanism that is controlled and coordinated centrally by our nervous system.
- The ability to cool down by sweating is believed to have played a key role in human evolution. Now known as the savannah hypothesis, this ability presumably gave humans the power to hunt animals by running them down to exhaustion in the scorching heat of the African savannah.
- Thin-film "wearable" microfluidic sweat sensors allow for monitoring metabolites present in your sweat.

10

The Flame of Microfluidics

The Capillary Action That Powers the Candle Wick and the Impact of Candles and Oil Lamps

In December 1860, seven hundred young people ranging from sixteen to twenty years of age sat shoulder to shoulder, waiting for the Christmas Lectures for Juveniles to begin. This lecture was part of a series of lectures held annually during the past 35 years. They knew they could count on this year's lecture on the candle to be as illuminating as lectures presented in previous years by the same eminent scientist. With its flame full of so much symbolism, the candle may seem a more appropriate subject for a poem than an illustrious scientific lecture. You may have been mesmerized by the cheerful flames smiling at a birthday girl from the top of her cake, or the icy candles weeping with churchgoers at a funeral in Italy. Perhaps you have experienced the awe-inspiring sight of thousands upon thousands of diya lamps in celebration of Diwali or the skyborne glowing swarm at night during the Yi Peng, Thailand's Sky Lantern Festival. Without realizing it, you were looking at one of the oldest, and perhaps most ingenious, microfluidic devices ever invented.

That year's lecturer—who had spearheaded the lecture series at the Royal Institution and had given more than half of them so far—was sixty-nine-year-old Michael Faraday, described by attendees as having "silvery hair" and "irresistible eloquence," an English scientist known in academic circles for his discoveries on electromagnetism and electrochemistry, among many other accomplishments in chemistry.[1] He had achieved international recognition and had even turned down a knighthood for his services to science on religious grounds, preferring to remain "plain Mr. Faraday to the end." His Christmas Lectures were undoubtedly a way to reconnect with his humble origins. The 1860 lecture was called "The Chemical History of a Candle," a lecture on the chemistry and physics of flames that was a favorite of his[a]—and of the public.

---

[a] Faraday had previously offered this lecture in 1848 and 1854.[1]
[1] Hammack, B., & DeCoste, D. *Michael Faraday's The Chemical History of a Candle.* Articulate Noise Books, 2016.

For that lecture, Faraday's majestic opening instantly captivated the students:

> There is not a law under which any part of this universe is governed which does not come into play and is touched upon in these phenomena. There is no better, there is no more open door by which you can enter into the study of natural philosophy than by considering the physical phenomena of a candle.[2]

With these words, Faraday proceeded to show a candle made of tallow, a type of animal fat. Those were the most affordable candles used in his day and also the most ancient ones.[3] As early as 3,000 BC, Egyptians soaked the dried pit of the rush plant in melted animal grease to make rushlights, an inexpensive candle. Although the Romans used oil lamps as their primary illumination source, they are credited with inventing wicked candles around 500 BC. They dipped a rolled papyrus (the wick) into tallow (the fat from sheep or cows) or beeswax. Another old use of candles is found in the Jewish Festival of Lights, or Hannukah, which centers on lighting nine symbolic candles and dates back to the second century BC. Every candle made since then works on the same principle, so Faraday knew the audience would have been familiar with candle wicks.

The wick of a candle or an oil lamp is a microfluidic device that is as ancient as it is ingenious. When the wick is first lit, the flame's heat melts the wax surrounding the wick, allowing for the liquified wax to climb up the wick by capillary action (Figure 10.1). In the case of an oil lamp, the melting step is not required because the oil is already in a liquid state. Still, the heat reduces the oil's viscosity, facilitating the capillary climbing of the fluid up the wick. In both candle and lamp, the flame burns the fuel (wax or oil) away on the outer tip of the wick, where there is abundant oxygen. The heat-producing chemical reaction between the fuel and oxygen produces (hot) carbon dioxide and water vapor, which rise in the air, leaving space for more fuel to wick in.[1] Hence, the wick—typically made of intertwined fibers—acts as a microfluidic pump that draws fuel until the candle or lamp burns out.

It would have been difficult to explain capillary action to a late-nineteenth-century classroom. Faraday's listeners did not have an advanced education, and the field of fluid mechanics was just being born. But Faraday was a gifted educator and mentally carried his young audience through two familiar experiments so they could grasp capillarity without any math:

> Capillary action conveys the fuel to the part where combustion goes on, and it is deposited there, not in a careless way, but very beautifully in the very midst

[2] Faraday, M. The chemical history of a candle. https://www.bartleby.com/lit-hub/hc/scientific-papers-physics-chemistry-astronomy-geology/the-chemical-history-of-a-candle (1860).

[3] Waters, P. Spermaceti candles. The Adverts 250 Project. https://adverts250project.org/2019/05/04/may-4-4 (2019, May 4).

Figure 10.1 The candlewick is an ancient microfluidic device. Using capillary action, the burning wick continuously pumps new fuel—the liquid wax melted by the flame's heat—up to the flame against gravity. The burning of the wax produces light (including heat), releases carbon dioxide and water vapor, and draws more molten wax up the wick.

Image source: User Jasmyn Favager on Unsplash.com. https://unsplash.com/photos/white-candle-in-black-background-QW538yvQ0Fo.

of the center of action which takes place around it. Now, I am going to give you two instances of capillary action. It is that kind of action or attraction which makes two things that do not dissolve in each other still hold together. When you wash your hands, you take a towel to wipe off the water; and it is by that kind of wetting, or that kind of attraction which makes the towel become wet with water, that the wick is made wet with the wax. If you throw the towel over the side of the basin, before long it will draw the water out of the basin like the wick draws the wax out of the candle.[2]

Not content with this account, Faraday came up with a clever "wick" made of a salt-filled tube to prove to the audience that capillary action could be implemented in a variety of scenarios:

Let me show you another application of the same principle. You see this hollow glass tube filled with table salt. I'll fill the dish with some alcohol colored with red food coloring. You see the fluid rising through the salt. There being no pores in the glass, the fluid cannot go in that direction, but must pass through its length. Already the fluid is at the top of the tube: Now I can light it and make it serve as a candle. The fluid has risen by the capillary action of the salt, just as it does through the wick in the candle.[1,b]

By the end of his impeccable explanations and flawless demonstrations, his captivated listeners must have felt they were experts in the matter.

One of the most ancient examples of wicked lamps is the Indian diya (Figure 10.2), still used by billions of people.[c] Diyas carry a sacred symbolism that is at the origin of Diwali, a several-day-long celebration of Prince Rama's return[d] by Hindus and embraced by others—over a billion people in India alone—during Kartika, a month in the Hindu calendar that typically overlaps October and November, during the time of the year when the nights

Figure 10.2 Diyas are rudimentary oil lamps made of clay and containing a wick, which brings oil to the flame by capillary action.
Image source: *Pxhere.com. https://pxhere.com/en/photo/1057976.*

---

[b] This text has been slightly adapted to modern English. For example, "capillary attraction" has been substituted by the modern term "capillary action."

[c] All the major religions of the Indian subcontinent—Hindu, Sikh, Buddhist, and Jain—believe that the warm, bright glow that emanates from a diya represents enlightenment, prosperity, knowledge, and wisdom, although its roots dig deep into the origins of Hinduism.

[d] Diyas carry a sacred symbolism that originates from the *Ramayana*, a Hindu epic from ancient India that narrates the life of Prince Rama in nearly 24,000 verses written in Sanskrit. The epic follows Rama's fourteen-year exile, his travels with his wife Sita and brother Lakshmana, and his triumphal return to his kingdom of Ayodhya after defeating the demon king Ravana.

grow longest and darkest. Diwali derives from the Sanskrit *Deepavali* or "row of lights." It is arguably India's most important festival of the year—much like Christmas in the West—including prayers, banquets, family gatherings, and fireworks. It is known as the "festival of lights" because people light candles or lamps in front of their homes, create *rangoli* patterns with chalk and flowers, place hundreds of diya in the middle of the sidewalk, launch thousands of floating oil lamps into the Ganges River, and set off fireworks on every street corner. In this way, India, once a year for many centuries, has celebrated "the victory of light over darkness, knowledge over ignorance, and good over evil"[4] with the help of microfluidics.

Diwali is not the only tradition celebrated by lighting candles and lamps. In China and nearby countries such as Taiwan, Thailand, and Japan, small hot air balloons called *sky lanterns* have been used in various festivals throughout history (Figure 10.3). Their invention is credited to third-century Chinese military strategist Kongming, who built a sky lantern to summon help when surrounded by enemy troops.[5] As the lamp heats the air inside, the lantern—still known as a Kongming lantern—slowly rises into the air due to the hot air's lower density than cold air. When the lamp extinguishes, the balloon gently descends as the air inside cools down. Similarly, *water lanterns*—typically square—are floated

Figure 10.3 Thousands of Khom Loi (sky lanterns) are floated into the air a few days ahead of the Yi Peng festival in Thailand.

Image source: User Takeaway on Wikipedia. https://en.m.wikipedia.org/wiki/Sky_lantern#/media/ File:Yi_peng_sky_lantern_festival_San_Sai_Thailand.jpg (CC BY-SA 3.0).

---

[4] Narayanan, V. Diwali: A holiday of goodness & light. In: *Celebrate Diwali with Sweets, Lights, and Fireworks* (ed. Heiligman, D.), 31. National Geographic Society, 2008.

[5] Deng, Y. *Ancient Chinese Inventions*. China Intercontinental Press, 2005.

in lakes and rivers throughout Asia during traditional festivals and ceremonies to worship the gods and welcome happiness. In all these awe-inspiring devices, light emanates from fuel that burns and ascends through a wick by the miracle of capillarity.

For centuries, different cultures throughout the world used candles for indoor illumination, using different materials for wax and wick.[6] The Chinese molded wax from insects in a paper tube and used rolled rice paper as the wick. In India, the wax was extracted from the fruit of the cinnamon tree. Tibetan people used yak butter to make their candles. Alaskan tribes learned to directly light the dried eulachon fish (also known as the "candlefish")—so oily that it burns like a candle when one end is lit. Due to the high availability of olive oil (used in oil lamps), candle-making remained relatively unknown in North Africa and the Middle East. After the fall of the Roman Empire, Europe experienced a severe disruption of the olive oil trade, and the more affordable tallow candles became more widely used. Throughout the Middle Ages, those who could afford the more expensive beeswax candles preferred them to tallow ones for their sweet smell and smokeless flame.[e]

Candles made from spermaceti wax—obtained from the crystallization of sperm whale oil—were introduced in the seventeenth and eighteenth centuries. They were a prized luxury item that produced odorless, smokeless, bright flames, even more expensive than beeswax candles.[7] Thus, those who could not afford them spent their dark evenings under the dim light and unpleasant smell of tallow candles. The spermaceti organ, which sperm whales appear to use for echolocation, is inside the forehead of the animal and can contain up to 1,900 liters of spermaceti oil.[8] It now seems cruel that an entire whaling industry was developed primarily to extract oil from a small part of the head of a sperm whale. The meat was discarded at sea. Spermaceti oil also found direct uses in oil lamps and as a lubricant for the new engines of the industrial age. The oils of

---

[e] The cost of candles conditioned the lives of people for a long time. Tallow candles were the most affordable choice but generated smoke and an acrid odor. Beeswax candles were expensive, and only the very rich and the clergy could afford them. When the first settlers arrived in New England, they found that they could extract wax from bayberries. Bayberry candles burned brightly and with a sweet smell, although it took 15 pounds of bayberries to obtain 1 pound of the green olive-looking wax. The chronicles of the time reflected that disparity. In the 1760s, the Duke of Newcastle paid £25.00 *every month* for beeswax candles to light his London house when by comparison the *yearly* wage of a housemaid at the time was approximately £3.00. Part of the allowance of servants would be in the form of a few candles—so they could see when they had to get up from bed or wanted to read in their room—with larger allowances for the more senior servants.[9]

[6] History of candle making. *Wikipedia*. https://en.wikipedia.org/wiki/History_of_candle_making (n.d.).

[7] Irwin, E. The spermaceti candle and the American whaling industry. *Historia Santiago* 21: 45–53 (2012).

[8] Spermaceti. *Wikipedia*. https://en.wikipedia.org/wiki/Spermaceti.

[9] Savage, W. The cost of 18th-century lighting. *Pen & Pension* [Blog] (2016, July 27).

Figure 10.4 This late-seventeenth-century whaling scene (titled *Walvisvangst*) was captured by Abraham Storck, a contemporary painter.
Image source: Wikipedia. https://en.wikipedia.org/wiki/History_of_whaling#/media/ File:Abraham_Storck_-_Walvisvangst.jpg. Public domain.

other whales were used occasionally but were less efficient and thus were three to five times less valuable. Whaling trips seeking spermaceti would last about three years and were fraught with dangers (Figure 10.4), as vividly immortalized in the famous epic tale *Moby-Dick*.[f] Tragically, over the nineteenth century and well into the twentieth century as well, hundreds of thousands of whales were butchered to the brink of extinction,[10] only so that a more pleasant light would illuminate the nights of the high society in the Western World.

Candle-based illumination lasted until the advent of electricity in the late nineteenth century. Candles were not easily affordable until 1834, when Joseph Morgan, a pewterer from Manchester, England, invented a machine that could mass-produce candles by molding melted wax at approximately 1,500 units per hour.[6] At about the same time, in 1830, German chemist Karl von Reichenbach developed the process for extracting paraffin wax from petroleum.[6] Paraffin wax burned with a bright flame without generating smoke or odors and

[f] American writer Herman Melville, the author of *Moby-Dick*, was inspired for his 1851 novel by a 21-meter-long albino whale of that period called Mocha Dick that proved exceptionally difficult to capture.
[10] NOAA Fisheries. Sperm whale. https://www.fisheries.noaa.gov/species/sperm-whale (n.d.).

was cheap to produce, so it instantly rendered spermaceti wax obsolete. Most modern-day candles are manufactured by molding paraffin wax. In 1882, the Edison Electric Illuminating Company of New York first brought electric light to parts of Manhattan. But electrification was slow. In the 1890s, the few cars and horse carriages equipped with headlights used oil lamps.[11] By 1925, half of the homes in the United States still did not have electric power—thus, they had to burn candles in the dark.[g] Despite the complete electrification of the United Kingdom in the 1970s, the miners' strikes sent a reminder that candles could still be essential. The miners' grievances snowballed into generalized strikes,[13] including the power station workers, that forced the UK government to limit commercial electricity consumption to three days a week starting January 1, 1974. During those days—for a restriction period that lasted seven long, cold weeks—the only illumination source was candles, and stores quickly ran out of them.[h,14]

The candle has now been mostly relegated to a celebratory or symbolic role during birthdays and other events, but we should be reminded that candles and their relatives have enlightened Humanity for several millennia. The candle and oil lamp wick, the ingenious capillarity-powered appliance so well described by Faraday, allowed the scribes of Antiquity and the Middle Ages to extend their working hours late into the evening to copy the codices and scrolls that were needed to pass on knowledge—law, history, medicine, math, technology, and science—from one generation to the next. Candles and oil lanterns provided more economical and safer alternatives to fires and torches, lighting up balls in royal halls and storytelling in family gatherings for centuries. Just as Faraday's wisely worded lectures helped expand the minds of young people every Christmas, the slow wicking of candles in homes throughout the world helped prolong a constellation of crafts, readings, conversations, parties, and banters into the dark hours of the night for thousands of years, contributing to the

[g] Even today, electrification is far from complete. As recently as 2020, 13% of the world's population (~940 million people) lacked adequate access to electricity.[12] Most of them are concentrated in Africa; in sub-Saharan countries such as South Sudan, Chad, and the Central African Republic (former colonies of the United Kingdom and France), the percentage of people with access to electricity is a meager 7–15%.[12]

[h] People worked three days a week during the seven-week-long restrictions. Television stations had to stop broadcasting at 10:30 p.m. every night, and people took baths in boiled water. The social unrest, known as the "Winter of Discontent," caused two government changes.[14]

[11] Admin, P. History of headlights: The story of the use of lasers from oil lamp in car headlight! https://www.prohori.com/english/history-of-headlights (2020, November 7).

[12] Ritchie, H., & Roser, M. Access to energy. *Our World in Data*. https://ourworldindata.org/energy-access (2020).

[13] Gibbs, E. In 1972, Britain's miners showed the power of the working class. *Jacobin*. https://jacobin.com/2022/02/1972-coal-miners-strike-fiftieth-anniversary-ncb-num (2022, February 22).

[14] Roller, S. When the lights went out in Britain: The story of the three day working week. *History Hit*. https://www.historyhit.com/when-the-lights-went-out-in-britain-the-story-of-the-three-day-working-week (2021, September 21).

transmission of knowledge and enhancement of productivity throughout the ages. Microfluidics has illuminated the expansion of our cultures and intellects.

Summary

- The candle wick is a very old microfluidic device that relies on capillary action to draw molten wax to the flame.
- Candles made from spermaceti—sperm whale oil—were introduced in the seventeenth and eighteenth centuries to produce odorless, smoke-less bright flames and were a luxury item.
- Capillarity-based illumination has contributed to the augmentation of knowledge by expanding the light hours into the night, but in many under-electrified regions it is also a sign of lack of industrial progress.

11

The Magic of Paper

The Capillary Action That Draws Ink out of Various Writing Devices into Paper

For reasons that remain a mystery, a remarkable man named Cai Lun underwent castration in China about nineteen centuries ago, allowing him to enter the prestigious corps of eunuchs at the imperial court (Figure 11.1). A eunuch is a person who has been castrated,[a] often with a specific social motive. Records show that, as a court eunuch, in 105 AD Cai Lun announced he had invented a papermaking process.[1] Every time we write or paint on paper, ink wicks by capillarity into the paper's fibers. In our digital age of touchscreens, we often do not think much of this microfluidic inking process nor realize how much we owe it to Cai Lun.

The invention of paper as a writing medium represented a monumental advance in recordkeeping and catalyzed a much-needed printing revolution. Since the abolition of feudalism in 221 BC by Qin Shi Huang, China's first Emperor and founder of the Qin dynasty, the government had been run by bureaucrats. Over time, these positions were filled through a system of civil

Figure 11.1 A 1962 Chinese postage stamp commemorating the 1,900[th] anniversary of Cai Lun's birth.

Image source: Wikipedia. https://www.roughtype.com/?p=4595. Public domain.

---

[a] Castration is the intentional removal of the testicles. In Imperial China, castration included emasculation (removal of the penis as well as the testicles) and was required for employment at the court.

[1] Needham, J. *Science & Civilisation in China, Vol. 5, Part I: Paper and Printing.* Cambridge University Press, 1985.

service examinations based on Confucius' books. All of these books had to be replicated for study. Upon appointment, these officials—named *mandarins* or "bossy" by the Portuguese because of their great zeal for their jobs—generated growing numbers of reports and letters that required copying. Furthermore, after the arrival and expansion of Buddhism into China, beginning in the first century AD, Buddhist monks started translating large amounts of Buddhist texts from Sanskrit into Chinese.[b] Yet ideograms were cumbersome to copy in large numbers. The only replication method known during Cai Lun's time, around the beginning of the second century AD, was the carved seals the Chinese had used since at least 1,000 BC to stamp their name in ink on materials such as silk. At the time, scribes would write on silk or rectangular bamboo slips that were joined like window blinds and stored in rolls. But silk was expensive, and the bamboo rolls were heavy. A productive day of reading by the emperor was reported as a *shi* (30 kilograms) worth of bamboo slips.[1] One day, Cai Lun realized that the paper wraps and clothing used in his native province could be handy if employed as a light, inexpensive support for writing.[2] The invention of paper earned him fame throughout the empire.

Cai Lun was born into a poor family in Luoyang in the Henan Province. His papermaking process utilized the bark of *ku* or paper mulberry (Figure 11.2).[2] This type of bark was being used in his native area—and in many places around the world—to wrap objects and to make non-woven clothes by beating and dissolving the bark fibers in water and drying them into thin sheets called *tapa*, a primitive and rough form of paper. They were uncomfortable, but the density of tapa made it an excellent insulating material for the winter months. Although tapa was considered a poor substitute for silk wraps, Cai Lun had likely seen packages of various goods wrapped in it in Luoyang. Unlike silk wraps, a person could scribble a name or note on the box. So, when he arrived at the court, he put his knowledge of *ku* to good use. He boiled the mulberry fibers to a pulp and beat it with a wooden or stone mallet. Then he placed a thin layer of the beaten pulp onto a wooden or cloth sieve to remove the excess water and let it dry. He had just invented paper.[2]

In the following centuries in China, Cai Lun's paper facilitated the duplication of the complex ideographic script and enabled the extension of seal printing to large-scale copying of texts.[2] Before the appearance of the printing press, people printed with chiseled stone slabs or steles. You can understand how the Chinese printed books and maps if you visit Xi'an. Xi'an was the capital of China

---

[b] The first Buddhist translators were An Shigao (second century AD) and Kumarajiva (fourth century AD).

[2] Tsien, T. Why paper and printing were invented first in China and used later in Europe. In: *Explorations in the history of science and technology in China.* Shanghai Chinese Classics Publishing House, 1982.

Figure 11.2 Cai Lun's paper-making process. Illustration from the *Tiangong kaiwu*, a compendium on industry, agriculture, and artisanry written in 1637 by the late Ming period scholar Song Yingxing.

Image source: Tiangong kaiwu. Public domain.

under twelve dynasties and the starting point of the famed Silk Road in ancient times. Most visitors go to Xi'an to admire the Terracotta Warriors—the earthenware army built to protect Qin Shi Huang in the afterlife—and his tomb. But in a nearby Confucian temple, starting in the Tang Dynasty (618–907), scribes carved stone steles so that people could replicate Confucian texts or paintings by inking the steles and rubbing paper rolls on them. Over the centuries, the collection grew to more than 3,000 steles. The so-called Stele Forest of Xi'an contains administrative or engineering texts (such as instructions for building a temple or repairing a canal), as well as poems and the works of great Chinese scholars. An incredibly accurate map of China from 1142 AD is also available to travelers. This was the photocopy center of ancient China.

Cai Lun's paper turned out to have the ideal properties for the absorption of fluids. Looking at a piece of paper under a powerful microscope, you will see a dense jungle of cellulose fibers leaving many pores in between (Figure 11.3). Cellulose is the fibrous material covering plant cells that is extracted during

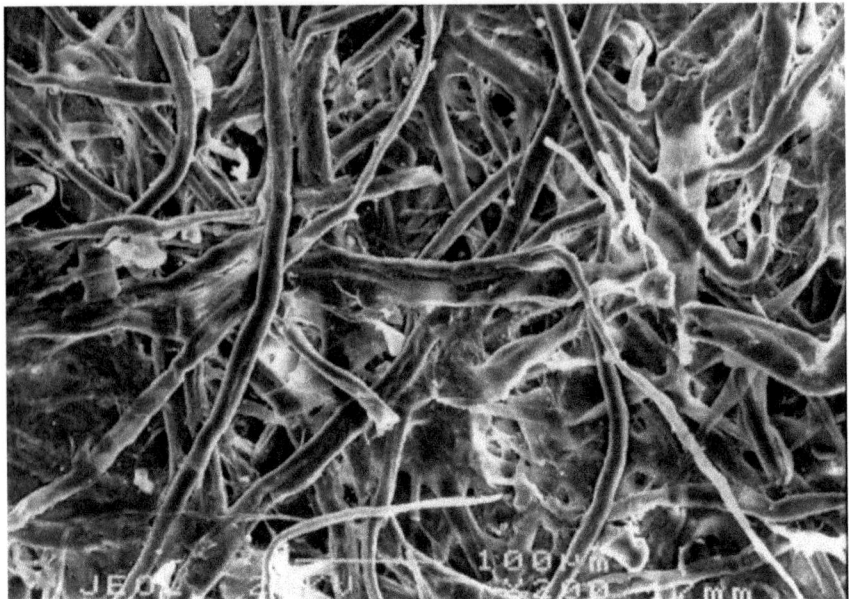

Figure 11.3 Scanning electron micrograph of a piece of paper. The whole image is approximately 0.6 millimeters wide, and the widest fibers are around 10–20 micrometers wide. When you place ink or paint on paper, the liquid wets the surface of the fibers and surface tension wicks it into every cavity.

Courtesy of Prof. Claire Davis and Chris Hardy, School of Metallurgy and Materials, University of Birmingham. https://www.flickr.com/photos/core-materials/4419088159/ (CC BY 2.0 Deed).

papermaking. Through the pores, water advances by capillary wicking, aided by cellulose's great affinity for water (or hydrophilicity). Each cellulose fiber loves to get wet in this tangle. Propelled by the hydrophilicity of cellulose at every corner, water has many opportunities to seep through the cracks. Hence, paper behaves like a magic microfluidic pump for all aqueous fluids.

Crucially, as Cai Lun wisely determined, the astounding absorptive properties of paper could be used not only for personal clothing and hygiene but also for transmitting knowledge. On a physical level, the transfer of an ink blotch to paper appears like that of a food stain or bodily fluid to cloth. The fluid spontaneously wicks into the spaces between the paper or the cloth fibers by capillarity. However, there is a critical difference between the ink transfer and the food stain. The ink is usually deposited in an intentional, not accidental, location (Figure 11.4). It carries information.

The capillarity-based transfer of information that Cai Lun introduced to the court did not consume any power and did not require any energy to maintain. Alas, there is no magic, but plenty of ingenuity. Of course, there is an energy

Figure 11.4 The transfer of ink to paper from a brush is pumped by capillary action.

Image source: Pxhere.com. https://pxhere.com/en/photo/1101239.

expenditure in ink printing, but the energy is stored *ahead* of the printing process *in the ink itself*—to be precise, *on the surface* of the ink, in the form of surface tension. Surface tension powers the transfer of ink onto the paper by capillarity. Ink printing on paper has been the predominant knowledge-transfer technology for two millennia, aided by the fact that it does not cost humans any energy. Surface tension and capillarity—two of the fundamental engines of microfluidics—have been doing all the work for us.

The West did not appreciate the absorbent properties of paper (including hygiene uses) until much later.[c] The Arabs learned the papermaking technique from Chinese prisoners in the eighth century AD but did not make much of it. Because the Roman alphabet was easier to copy than Chinese cuneiform, the demand for books could be met with manual copying by scribes on other supports such as wood and leather. Hence, paper did not start circulating within

---

[c] Other applications of paper as a porous medium underwent a similar pattern of delayed adoption by the West. In the sixth century AD, the Chinese already used paper for hygiene. By the Middle Ages in Europe, China was consuming toilet paper sheets by the millions. Westerners, on the other hand, scorned the use of toilet paper. In 851, an Arab wrote, "The Chinese are not very careful with their hygiene and do not wash up after going to the bathroom. They only clean themselves with paper."[3]

[3] Temple, R. *The Genius of China: 3000 Years of Science, Discovery and Invention*. Touchstone Books, 1989.

Europe until the eleventh century, and large-scale printing did not begin until the fifteenth century with Gutenberg's moveable printing press.[3]

The development of the engraving technique by German artist Albrecht Dürer in the fifteenth century and the etching technique by Dutch painter Rembrandt in the seventeenth century allowed for the incorporation of image reproductions alongside printed text. In both techniques, the delicate features of a drawing were inscribed on a plate and transferred to paper by wicking after inking the plate. In his time, Rembrandt was primarily known for his etchings, and paper reproductions of his work circulated throughout Europe. Thanks to Gutenberg's printing press, books and political and religious pamphlets had been mass-produced and distributed across Europe for almost 200 years, but they generally lacked images. In 1633, a French contemporary of Rembrandt, Jacques Callot, produced "*Les Misères et les Malheurs de la Guerre*," a series of eighteen powerful and influential etchings that were widely reproduced in newspapers and magazines of the time and contributed to raising people's awareness of the cruelty of the Thirty Years' War (Figure 11.5). Through etchings and printing, capillarity powered the reproduction of words and images onto paper from the fifteenth century onward, later supplemented by the invention of photography in the 1830s.

Humans developed a few absorptive materials besides paper, most notably papyrus and parchment. This variety is not surprising because at least four ancient civilizations—Sumer, Egypt, China, and Mesoamerican cultures—may

Figure 11.5 Engraving "*Le Pillage*" (Plate 5: Pillaging a house), from the series of eighteen engravings *Les Misères et les Malheurs de la Guerre* by Jacques Callot.

Image source: Wikipedia. https://en.wikipedia.org/wiki/ Les_Grandes_Mis%C3%A8res_de_la_guerre#/media/ File:Pillaging_from_The_Miseries_and_Misfortunes_of_War_by_Jacques_Callot.jpg. Public domain.

have independently developed writing between 3,000 and 5,000 years ago.[4] Papyrus—from which the word "paper" is derived—is a porous material extracted from the interior of reed stems.[d] Reeds have strong stems that grow to several meters tall. They were very abundant in the marshy delta of the Nile River thousands of years ago.[e] Parchment has been in use since at least 1,500 BC and is prepared by scraping and drying animal skin.[f]

The superior properties of cellulose remain a key element of why we use paper today. Most of the paper in the world is now harvested from wood, mainly from tree trunks, which contain the minimum number of nonfibrous elements.[g] Paper towels, toilet paper, paper napkins, and tissue paper—sanitary advances that did not reach the West until the twentieth century—are all made of the same molecular material, cellulose, and use the same physical principle, capillary wicking, to pump fluids away from an undesired location and into the paper's fibers (Figure 11.6). Paper manufacturers reuse cotton rags and garments to make high-grade, finer-fiber paper for banknotes, legal documents, bond letterheads, and so on. As of 2020, the global paper market has reached $350 billion. More than a tenth of this market is in sanitary paper; slightly more than half of it is taken up by wrapping and packaging, and around one-fourth is taken up by printing.[5]

Perhaps[6] you are not the type to thank microfluidics whenever you use the toilet or wipe off the kitchen counter with a paper towel. But even in our screen-ruled world, we still owe to capillary action the ink motifs in every bill

[d] Reed is a common denomination for various types of grass-like plants that live in swamps.

[e] The Egyptians used the reed variety known as Nile grass (*Cyperus papyrus*) to weave ropes, baskets, blankets, mats, and sails. To make a sheet of papyrus, the Egyptians split open the central portion of the stalks and laid many strips next to each other in parallel onto a board with the pulp facing up. Next, they poured a thin paste of flour, vinegar, and water onto the strips and covered them with a second layer of strips at right angles with the pulp facing down. Finally, they pressed or hammered the two layers together and let them dry to achieve a sheet of papyrus.

[f] The process is very tedious and the throughput is impractically low—it has been estimated that approximately 300 sheep are needed for a book the size of a Bible. The word "parchment" is derived from the Greek city of Pergamon in Asia Minor. King Eumenes II (197–150 BC) from Pergamon enviously admired the culture of Egyptian Ptolemaic kings and their Alexandria library, so he founded the Pergamon library, which contained 200,000 volumes at some point. In response to this new interlibrary rivalry, King Ptolemy VI of Egypt forbade the export of papyrus, so Pergamon became a thriving center of production of parchment as an alternative writing material—hence, the origin of the name.

[g] Approximately 35-40% of the harvested wood goes to paper production. The majority of paper is made from pine trees. Some sources estimate that a standard pine tree produces about 10,000 sheets of paper or twenty reams of paper (each ream is 500 sheets).[6]

[4] Robinson, A. *The Story of Writing*. Thames and Hudson, 1999.

[5] Fortune Business Insights. Pulp and paper market size & share. https://www.fortune businessinsights.com/pulp-and-paper-market-103447 (2021).

[6] Ribble. How much paper comes from one tree? https://ribble-pack.co.uk/blog/much-paper-comes-one-tree (n.d.).

Figure 11.6 The absorption of fluids by a paper napkin is powered by capillarity.

Image source: kzww/Shutterstock. https://www.shutterstock.com/image-photo/tomato-stains-on-white-paper-1919436215.

of paper currency that has been in our hands; every label you read on all those cans you buy and packages you receive; every printed book, newspaper, and magazine you have ever shared with a loved one (see Chapter 6); and the colorful paper that wraps wonderful surprises around your birthday and holiday gifts.

* * *

When it comes to ink and writing, microfluidics is in the give and take. Throughout the ages, various tools were developed for transferring ink to these absorptive materials—the reed, the quill, the fountain pen, and the brush. In all cases, wicking acts both within the tool and within the paper. The capillary action of the tool loads the ink into the device, and the capillary action of the paper unloads the ink from the tool into the paper.

The oldest ink-dispensing tool known is the reed pen. A 4,500-year-old statuette of an Egyptian scribe shows him holding a papyrus and a reed pen.[4] Reeds have strong stems that grow to several meters tall. The Egyptians may have adapted a Sumerian etching tool to deliver ink. In ancient Egyptian sites dating from the fourth century BC, archeologists have found reed pens with a split nib used by Sumerians to make their cuneiform inscriptions in clay like those used in modern fountain pens. As the tip of the reed pen was dipped in ink, ink rose via the split into the hollow shaft of the pen by capillarity. The Greeks

Figure 11.7 The work of a student of Arabic calligraphy, using bamboo pens ("qalams") and brown ink, tracing over the teacher's work in black ink.

Image source: User Aieman Khimji on Wikipedia. https://en.wikipedia.org/wiki/ Islamic_calligraphy#/media/File:Learning_Arabic_calligraphy.jpg (CC BY 2.0).

and the Romans also used similar reed pens. Bamboo reed pens—called *qalam* (from the Greek *kálamos*, "reed")—are still widely used for Islamic calligraphy (Figure 11.7).

Reed pens, however, are stiff and do not retain a sharp point for very long. More durable *quill pens* (Figure 11.8) superseded reed pens sometime before the sixth century.[h] Quills were made by cutting the end of feathers from a large bird such as a goose, a swan, or a turkey. As the writer dipped the quill's tip in ink, ink rose into the feather's hollow shaft (called the calamus) by capillarity. People used quill pens to write and sign historical documents such as the Magna Carta in the thirteenth century all the way to the American Declaration of Independence in the nineteenth century. Even now, each day the U.S. Supreme Court is in session, twenty goose quill pens are placed on the tables.[7]

---

[h] At the beginning of the seventh century, the Archbishop of Seville wrote of quills being used for more than 100 years.[7]

[7] History of Pencils. History of quill pens. http://www.historyofpencils.com/writing-instruments-history/history-of-quill-pens (2023).

Figure 11.8 A quill pen.

Image source: Pxhere.com. https://pxhere. com/en/photo/1601560.

Capillary action also works to empty the quill during writing. When the writer gently presses the tip of the quill into paper, the ink enters the paper's microscopic mesh of cellulose, which—aided by gravity—acts as a capillary pump to pull the ink out of the pen's calamus. Modern *fountain pens*, which incorporate an ink reservoir, utilize a metal nib with a sub-millimeter-sized slit that acts as a capillary channel between the reservoir and the paper (Figure 11.9).[i]

In East Asia, writing became an art form and developed in the form of ideograms featuring complex shapes that were—and still are—painted with an ink brush. The process is the same as when painters paint with water-based inks, such as watercolor or acrylic paint (Figure 11.10). Since ancient times, brushes have been made with animal hair. Brushes made with hog bristle absorb paint very well. Many other animals have been used, such as the hair from sables (a small mammal from Siberia), squirrels, badgers, horses, oxen, goats, pigs, deer, buffaloes, mice, rabbits, wolves, tigers, and even people. Nowadays, they

---

[i] Intriguingly, the manuscripts of Leonardo da Vinci, who lived in the fifteenth century, do not have the unequivocal, alternately fading trace typical of quills. This finding has led scholars to the conclusion that the genius artist and engineer must have invented a precursor of the fountain pen.

Figure 11.9 A fountain pen.
Image source: Pxhere.com. https://pxhere.com/en/photo/1506409.

are also made of hydrophilic polymers (usually nylon or polyethylene) to optimize the wetting. As the writer or artist dips the brush in the ink, the ink wicks into the spaces between the hairs by capillarity, and surface tension retains the ink within those spaces when the artist or writer retrieves the brush from the ink. As with the quill, as soon as the brush contacts the paper, capillary action fills the fibers of the paper, which also pumps the liquid out of the brush. Most people must have used a brush to write or paint at least once in their lifetime. In 2019, the market for paint brushes generated more than $2 billion, with more than 400 million brushes sold globally. This battery-free mechanism we call capillary action has been powering two quintessentially human activities—writing and painting—for millennia.

Even if you do not own any brushes or have not heard of microfluidics, you have probably owned a few dozen *ballpoint pens*. The ballpoint pen (Figure 11.11, left) is an ingenious microfluidic device that uses the principle of surface tension to spread ink on a tiny, rolling metal ball at its tip. The ball is

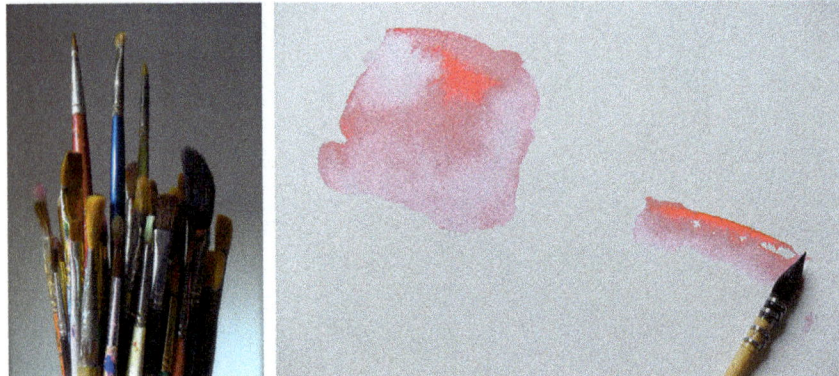

Figure 11.10 In watercolor painting, capillary action is the driving force that wicks the ink into the brush fibers, loading the brush. Capillary action is also responsible for ink wicking from the brush into the paper fibers, producing the colored smears.

Image sources: (Left) Pxhere.com. https://pxhere.com/en/photo/978436. (Right) Pxhere.com. https://pxhere.com/en/photo/1552539.

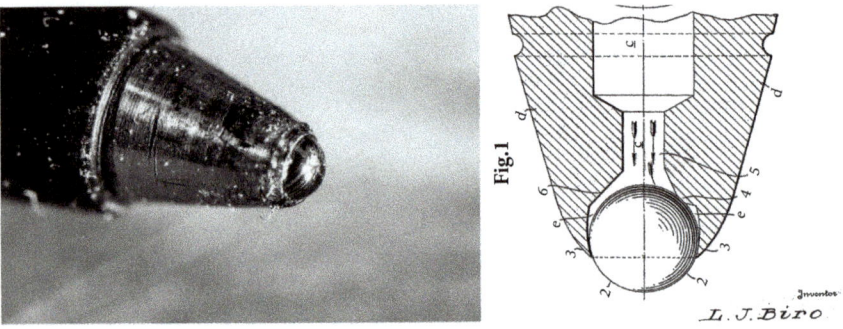

Figure 11.11 (Left) A ballpoint pen. (Right) Cross-section schematic of a ballpoint pen's tip, from Bíró's patent.

Image sources: (Left) user The Manic Macrographer on Pxhere.com. https://pxhere.com/en/photo/290115. (Right) Google Patents. https://patents.google.com/patent/US2390636A/en.

made of a hard metal such as steel, brass, or tungsten carbide and ranges from 0.5 to 1.2 millimeters in diameter. A socket with a round opening holds the ball in a snug fit, plugging the bottom of the ink reservoir inside the pen's body (Figure 11.11, right).

This plug is the key. The ball and the socket do not form a perfect match—there are some imperfections on both metal surfaces due to their microscale roughness. Why does the pen not leak through the space left by these microscopic irregularities? The ink has a high viscosity, so it cannot easily flow through the small gaps. If the ink were too watery, or the irregularities larger, the plug would not work, and ballpoint pens would leak. You might have experienced the "bursting" of a ballpoint pen on a very hot day: when the ink is heated, its viscosity decreases, allowing it to leak through the gaps of the ballpoint. Under normal conditions, however, the only way for ink to make it to the outer side of the pen is when the ball rolls. The rolling action covers the ball in a microscopically thin layer of ink and, at the same time, creates a smooth glide over the surface as the writer scribbles on paper. Although the ink has a high viscosity, it also likes to wet the ball's metal surface—that is, it has a low surface tension in contact with the metal. As a result, the ball's irregularities become all bathed in ink. This stable, microfluidic film makes it past the socket's opening when the ball rolls, thick enough to deposit the ink on an absorbent paper.

Ballpoint pens are manufactured and sold by the billions, making them the most successful microfluidic device in history. The first ballpoint pen was commercialized in the 1940s in Argentina by Nazi exiles, Hungarian newspaper editor László Bíró and his brother Győrgy, a chemist. Bíró had become exasperated with the poor reliability of fountain pens, which caused blotches of ink on his shirts without warning and made those hair-raising scratchy sounds. Left-handed writers smudged their fountain pen scribbles as the hand slid over the wet ink. Bíró was inspired by the ink used in newspaper printing presses, which dried quickly and did not smudge, and by an expired 1888 patent by American inventor John Loud that described a pen with a large metal ball at the point for marking leather.[8]

Bíró understood that he needed a viscous ink coupled with a (much smaller) ball-socket mechanism, allowing controlled flow on a microscopic scale. Unlike previous water-based inks, Bíró's ink was composed of dyes suspended in an oily solvent, which formed a thick fluid that dried quickly and helped lubricate the ball tip due to the ink's low surface tension. It was a success. His patent, filed in 1938, did not go unnoticed. *The Wall Street Journal* reported that "a simple but remarkable invention came into a world about to be convulsed by death and destruction."[9] One of the first big orders came from the Royal Air Force during World War II. The pilots, who needed to write at high altitudes, preferred the leak-proof Bíró pens to the fountain pens that became leaky at low pressures and caused messy stains on maps and uniforms.

[8] Bellis, M. Inventor Laszlo Biro and the battle of the ballpoint pens. ThoughtCo. https://www.thoughtco.com/ballpoint-pens-laszlo-biro-4078959 (2018, February 22).

[9] Smith, N. Late great engineers: László Bíró—mightier than the sword. *The Engineer*. https://www.theengineer.co.uk/content/in-depth/late-great-engineers-laszlo-biro-mightier-than-the-sword (2022, February 3).

Amid the competition of many Bíró pen copycats of lower performance,[8] French businessman Marcel Bich, who owned a fountain pen company in Paris, licensed the ballpoint pen patent from Bíró and renamed the company BiC. BiC re-engineered Bíró's ballpoint pen to lower the manufacturing costs and started mass-producing the ubiquitous BiC Cristal ballpoint pen in 1950. BiC now sells approximately 14 billion pens every year and has sold more than 100 billion pens since 1950.[10]

BiC pens—still known in Argentina as "biromes"[j]—are found in all corners of the world. A standard ballpoint pen can produce a line of about 2 km, or about 50,000 words—100 pages of text. In the United States alone, each person uses more than four ballpoint pens on average annually. Close to 125 ballpoint pens are sold every second, amounting to nearly 4 billion pens sold yearly.[10] By comparison, there are less than 7 billion smartphones worldwide (less than half of which are replaced annually), 2 billion computers, and 1 billion cars circulating today, and most humans do not own a phone, a computer, or a car—so in a count of devices sold, microfluidics beats microelectronics by far. Not even Cai Lun could have imagined that a simple pen used on "his" paper would ink the essays and maths of billions of schoolchildren, the birthday cards and letters of myriad friends, and the signatures of contracts and checks of everyone else.

Summary

- The invention of paper as a writing surface is attributed to Cai Lun, a Chinese eunuch in the imperial court about nineteen centuries ago.
- Ink loads into brushes by capillarity and unloads or wicks into the fibers of the writing medium—papyrus, parchment, or paper—also by capillarity.
- The energy required to pump the ink (in and out of the brush and into the writing medium) is stored on the fluid's surface in the form of surface tension.
- Ballpoint pens are microfluidic devices that rely on the surface tension of a viscous ink to deliver the ink through a rolling ball socket.

---

[j] The Bíró brothers and business partner Juan Jorge Meyne sold their pen in Argentina under the name Birome (a fusion of Bíró and Meyne). In 1945, Eversharp Company acquired the rights to the pen and sold it under the name Eversharp CA ("CA" stood for "capillary action"), but it did not survive the competition of BiC.[8]

[10] Each second of every day, more than 125 ballpoint pens are sold. *South Florida Reporter*. https://southfloridareporter.com/each-second-of-every-day-more-than-125-ballpoint-pens-are-sold (2023, June 10).

- The first ballpoint pen was commercialized in the 1940s in Argentina by Hungarian brothers László and Győrgy Bíró. French businessman Marcel Bich licensed the patent from Bíró and renamed his company BiC, which started mass-producing the ubiquitous BiC Cristal ballpoint pen in 1950.
- At more than 100 billion units sold since 1950, the ballpoint pen is—by far—the most universally distributed microfluidic device in modern times.

12

The Man Who Wore a Sanitary Pad

The Revolution in Women's Health Care Brought About by Inexpensive Pads and Home Pregnancy Tests, Both Powered by Microfluidics

In 1998, as a twenty-nine-year-old inexperienced newlywed in the town of Coimbatore, located in India's southern state of Tamil Nadu, Mr. Arunachalam Muruganantham (Figure 12.1, top) was surprised to learn for the first time about his wife Shanti's monthly period. Stumbling upon Shanti using dirty rags as sanitary pads—"so dirty I would not use them to clean my bicycle," he explained[1]—he went to his neighborhood store to buy her a clean cotton pad. Muruganantham was surprised to get a clearly overpriced item made of simple cotton. Oddly enough, the sales clerk who sold the pad to him quickly wrapped it in newspaper as if he were selling a forbidden or shameful item. Without having ever set foot in a university, Muruganantham ventured on an odyssey during which he not only learned to manufacture and distribute sanitary pads efficiently and at low cost but also mastered the underlying principles of how and why some materials wick better than others, making them more or less suited to his enterprise. His unstoppable determination helped his wife—and many other women—have a more hygienic, comfortable, cost-effective, and, ultimately, more private way of managing menstruation.

In India, as in many places, menstruation has long been taboo. Large sectors of the Indian population still believe that menstruating women are impure. During their period, many women are often excluded from social, educational, and religious activities.[a] As recently as February 2020, sixty-eight young female undergraduate students at Shree Sahajanand Girls Institute of Gujarat, a western Indian state, were taken to the toilet. There, they were asked to individually remove their underwear and show it to their female teachers to prove that

---

[a] These exclusions happen in other cultures as well. For example, the book of Leviticus in the Old Testament of the Bible says, "When a woman has her monthly period, she remains unclean for seven days. Anyone who touches her is unclean until evening. Anything on which she sits or lies during her monthly period is unclean. If a man has sexual intercourse with her during her period, he is contaminated by her impurity and remains unclean for seven days, and any bed on which he lies is unclean" (Lev. 15:9–33).

[1] TED. Arunachalam Muruganantham: How I started a sanitary napkin revolution! YouTube. https://www.youtube.com/watch?v=zkQL7UJYDIY&t=334s (2012).

Figure 12.1 (Top) Mr. Muruganantham at the workshop where his machines are fabricated. (Bottom) A woman operating one of his machines. A press simultaneously compresses cellulose into three pads.
Source of the images: YouTube. https://www.youtube.com/watch?v=-bVkk3TcEb8. Fair use.

they were not menstruating. Someone had complained that menstruating students were not following "the rules." These rules for menstruating girls included sitting away from others at mealtimes, cleaning their dishes, and sitting on the last bench in the classroom. Further adding to this stigma, because commercially manufactured sanitary products have been cost-prohibitive, most women in rural India do not wear pads or tampons during their period and resort to unsanitary absorbent materials such as rags, leaves, sand, ash, wood shavings, newspapers, or hay. Similarly, in many places in Africa, women resort to unhygienic solutions because sanitary pads cost more than their average income ($1.25 per day).[2]

There is hope for such low-resource settings, thanks to Muruganantham's incredible persistence and inventiveness. His triumphant success, in the end, came only after a great deal of pain. Muruganantham grew up in a poverty-stricken family. After his father died in a traffic accident, his mother worked as a farm laborer. At fourteen years of age, he dropped out of school and took up various jobs to help at home. Marrying Shanti in an arranged marriage was one of the highlights of his life, and he felt he needed to do something good for her by solving her unhygienic pad problem. Upon returning home from the store that day in 1998 with the overpriced pad, Muruganantham decided he could do much better and make the pads himself. But he needed volunteers to test them. Because his wife and sisters were too embarrassed to cooperate—they were utterly mortified and thought he had lost his mind—he sought help from local medical students. They were as unenthused as the members of his family, which left only one test subject he could think of—himself. Muruganantham decided to use a portable container filled with animal blood and wear a cotton pad around town. "The first man to wear a sanitary pad," he would joke later.[1] Fellow villagers scorned him, and after eighteen months, his wife, ashamed by her husband's inexplicable obsession, left him. Still, he persevered. He finally managed to collect used sanitary pads from student volunteers, and he laid the pads down in his backyard to study them. This was too much for his mother, who was living with him then, and she left his house to live elsewhere. At some point, he felt in danger of being attacked by fellow villagers, so he left his hometown.

Through his study of various pads, Muruganantham realized they were not all created equal and that the best, commercially manufactured ones had a secret. Muruganantham was determined to find out what that secret was. After many letters and phone calls—a process that took several years of his life—a manufacturer finally contacted him and inquired about his "plant." At first, Muruganantham thought the manufacturer was referring to his garden. After clearing up the misunderstanding, he convinced the manufacturer to send samples of the raw materials used in its sanitary pads. When the cellulose boards from tree bark arrived, Muruganantham was exultant to realize he could get the same ones in the jungle nearby. After learning that the machines that process the cellulose into pads cost half a million dollars each, he devised a do-it-yourself type of machine (Figure 12.1, bottom). At a cost of only $860, and with some manual input, it grinds, de-fibrates, presses the cellulose into pads, and sterilizes the pads under ultraviolet light.

---

[2] Papyrus sanitary pad. *Wikipedia.* https://en.wikipedia.org/wiki/Papyrus_sanitary_pad (2006).

In 2006, Muruganantham visited the Indian Institute of Technology Madras (IIT) to seek advice. Unbeknownst to him, the IIT entered his invention for a national competition award, and it won first place out of 943 entries. Pratibha Patil, then President of India, handed him the award. He appeared on national television. And his wife—"*my* Shanti," as Muruganantham refers to her in interviews, a charming play on words since *shanti* in Hindi means "inner peace"—called to ask if she could return home. He could not have been happier.

Three things make Muruganantham a most remarkable individual. First, he did not rest until he understood why the materials used for commercial pads—cotton and cellulose—work so well, reverse-engineering the sanitary pads. His ingenuity, intellect, and drive overcame the obstacles of not having access to a science degree. At some point, he intuitively realized that the pads' fibers wick fluids by the same mechanism used by the porous cellulose fibers in paper to wick ink (see Figure 11.3). Like most inventors, Muruganantham drew on his daily experience, as we have been surrounded since childhood by materials and devices that wick fluids (Figure 12.2). Most fibrous materials, such as clothes, rags, cotton, fluff, pulp, mops, and sponges, absorb liquids when dry by capillarity (and some, such as sponges and mops, also even when wet by mechanical pumping) and have found many sanitary applications. Diapers, tampons, surgical pads, bandages, gauzes, Band-Aids, and wound dressings are all made with absorptive layers. These layers sometimes contain superabsorbent polymers that can absorb more than 400 times their weight in water. In low-income areas of western Uganda, papyrus—due to its high abundance—has been utilized to make sanitary pads that cost one-fourth the price of a conventional pad.[2] The papyrus sanitary pads, called MakaPads, are 100% biodegradable, and their manual production employs many. Like a one-person bioengineering team and entrepreneurial superstar, Muruganantham single-handedly grasped the complex process of pad-making only by looking at the pads, the boards of cellulose, and the capillary-wicking properties of both.

Second, one of the most extraordinary things about Muruganantham is that he does not seek profits from his invention. His reasoning has the clarity of a philosopher warning us against the perils of greed. "If you run after money, life has no beauty—it has boredom," he avowed in front of a mesmerized TED talk audience.[1] He had the opportunity to sell his machine to a commercial manufacturer, but he refused. "There are lots of people making billions of money [sic] and running for philanthropy afterward—why not start philanthropy on day one?" His goal from the onset has been to start a sanitary revolution, no less.[1] He chose to sell his machine to underprivileged communities in rural India, providing them with a low-cost manufacturing facility for hygienic pads and a consistent livelihood. He has designed his machines to be easy to use. The women make the pads and sell them for a small profit to other local women.

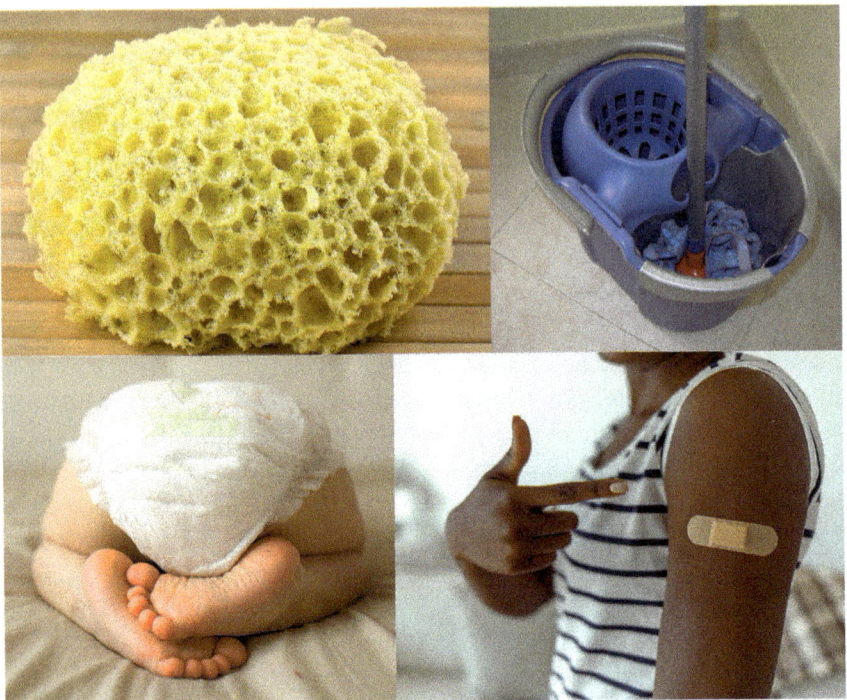

Figure 12.2 Common daily materials and devices that absorb fluids by capillary action. (Top left) A sponge. (Top right) A mop. (Bottom left) A baby diaper. (Bottom right) A Band-Aid.

Image sources: (Top left) Pxhere.com. https://pxhere.com/en/photo/652100. (Top right) User Husond on Wikipedia. https://commons.wikimedia.org/wiki/File:Janitor%27s_bucket_with_mop. jpg (CC BY-SA 3.0 Deed). (Bottom left) Denis Kalinichenko/Shutterstock. https://www.shutterstock. com/image-photo/small-child-sleeping-diaper-1025743165. (Bottom right) Jelena Stanojkovic/Shutterstock. https://www.shutterstock.com/image-photo/young-african-american-woman-showing-arm-213430015.

"I'm like a butterfly, flitting about, pollinating," he said. By empowering the women, he also fights the taboo. In 2007, a national survey found that only close to 24% of adolescent girls used sanitary pads. Five years later, 74% said they were using them. His do-it-yourself, low-cost machines have been installed in 4,800 points in rural India and twenty-nine other countries, earning him international recognition. He was featured in the 2018 Academy Award documentary *Period. End of Sentence.* Now known to the world as "Padman," the title of the 2018 Hindi film about him, he was included in *TIME* magazine's 2014 list of 100 Most Influential People in the World.

And finally, Muruganantham's genuine warmth and generosity are an inspiration. He does not hold any grudges, and he is contentedly back in his hometown with his Shanti, daughter, mother, and fellow villagers, who are no longer embarrassed about menstruation and this crazy "Padman." They now ask to take selfies with him.

* * *

Microfluidics underlies much of women's fight for dignity and independence. It is not just the sanitary pad: Two other devices also essential to the women's liberation movement—the tampon and the home pregnancy test—rely on capillary wicking as well. The simplicity of capillary action has empowered women to become individuals who manage their own health and sexuality.[b] Just as revolutionary for women as Muruganantham's inexpensive sanitary pads in India was the invention of the home pregnancy test in the 1960s. That invention led to the development of Clearblue, the much more convenient microfluidic pregnancy test based on capillary action.

Women have been seeking pregnancy tests for thousands of years. According to a 3,500-year-old Egyptian papyrus,[c] a doctor gave a recipe for predicting pregnancies using plants:

> Let the woman water [them] with her urine every day with dates [and] the sand, in two bags. If they [both] grow, she will bear. If the barley grows, it means a male child. If the wheat grows, it means a female child. If both do not grow, she will not bear at all.[4]

The Egyptian method was proven in the 1960s to be 70% accurate for predicting pregnancy, although it did not predict the child's sex. "Piss prophets" in the Middle Ages claimed to diagnose pregnancy by visually examining the turbidity or other features of a woman's urine. None of those claims have been supported, but there is indeed something in the urine of pregnant women: high levels of a hormone called *human chorionic gonadotropin* (hCG) that helps maintain pregnancy.

---

[b] Not all devices powered by capillarity have directly benefitted women's emancipation. It could easily be argued that devices such as the mop and the rag—which also rely on wicking to perform their water-collecting function—have often been used to subjugate women to cleaning tasks.

[c] Several Egyptian texts include medical advice on conception, fertility, pregnancy, and/or birth control (among others), including the Kahun Gynaecological Papyrus (c. 1800 BC), the London Medical Papyrus (c. 1782–1570 BC), the Ebers Papyrus (c. 1550 BC), and the Berlin Medical Papyrus (also known as the Brugsch Papyrus, c. 1570–1069 BC).[3]

[3] Mark, J. J. Egyptian medical treatments. *World History Encyclopedia*. https://www.worldhistory.org/article/51/egyptian-medical-treatments (2017, February 20).

[4] Ghalioungui, P., Khalil, S., & Ammar, A. R. On an ancient Egyptian method of diagnosing pregnancy and determining foetal sex. *Med. Hist.* 7: 241–246 (1963).

The first home pregnancy test came from a woman's insight, not surprisingly. In 1967, Organon Pharmaceuticals, a company in New Jersey, hired Margaret Crane as a freelance graphic designer to work on a new line of cosmetics. In the lab, she saw some test tubes and was told that they were for a pregnancy test, the so-called *hCG immunoassay.*[5] The immunoassay technique,[6] developed only a few years before, used antibodies as "molecular sensors" to detect a biomolecule of choice—in this case, hCG. What caught Crane's attention was the possibility of detecting pregnancy by performing a simple color-changing chemical reaction in a test tube with a woman's urine.

"Could a woman use this technology at home?" Crane wondered. At the time, pregnancy tests had been performed for decades in animals such as rabbits[7,d] and frogs,[8] which were not something a woman could easily do at home. To prove her point, Crane designed a set of test tubes that would hold the necessary solutions and be easy to use—privately—at home. The kit included a paper clip holder and a mirror. "I was absolutely certain this product would be very useful. That a woman should have the right to be the first to know if she was pregnant and not have to wait weeks for an answer," Crane would say later.[9] Organon initially hesitated but finally decided to patent and market Crane's product. The product, named the Predictor, cost $10 (around $75 today) and was sold first in Canada[e] in 1971.

By then, a company called Unilever was already developing a dip-stick version of Crane's immunoassay.[10] Their new Medical Division, seeking new markets for expansion and diversification,[f] had recruited immunologist Philip Porter to strengthen its science base in the 1960s. Porter had a PhD in immunochemistry

---

[d] Hence, the euphemistic expression "the rabbit died" when the test turned out positive, even though the rabbit did not really die.

[e] It was not sold in the United States until 1977 due to sexual morality concerns and (male) doctors arguing that women might not be able to cope with the results if a doctor was not present.[11]

[f] In the 1950s, Unilever was a vast conglomerate of companies with products ranging from detergents to fish fingers.

[5] Yong, E. How a frog became the first mainstream pregnancy test. *The Atlantic.* https://www. theatlantic.com/science/archive/2017/05/how-a-frog-became-the-first-mainstream-pregnancy-test/525285 (2017, May 4).

[6] Yalow, R. S., & Berson, S. A. Assay of plasma insulin in human subjects by immunological methods. *Nature* 184: 1648–1649 (1959).

[7] Olszynko-Gryn, J. The demand for pregnancy testing: The Aschheim–Zondek reaction, diagnostic versatility, and laboratory services in 1930s Britain. *Stud. Hist. Philos. Sci. Part C Stud. Hist. Philos. Biol. Biomed. Sci.* 47: 233–247 (2014).

[8] Shapiro, H. A., & Zwarenstein, H. A rapid test for pregnancy on *Xenopus laevis. Nature* 133: 762 (1934).

[9] Bonhams. Crane, Margaret, inventor—the first home pregnancy test. https://www.bonhams. com/auctions/22407/lot/37 (2015, June 16).

[10] Jones, G., & Kraft, A. Corporate venturing: The origins of Unilever's pregnancy test. *Bus. Hist.* 46: 100–122 (2004).

[11] Romm, C. Before there were home pregnancy tests. *The Atlantic* (2015).

and must have known that many researchers before him had used paper strips to separate and analyze chemicals cheaply.[4,12-17] Porter established a stellar research team and program at Unilever and developed the ground-breaking "dip-stick" concept of a simple, efficient one-step immunoassay under the subsidiary Unipath.

Unilever introduced the universal two-stripe pregnancy test, Clearblue One Step, in 1988. This and similar products are still widely used today, relatively unchanged. Inside the pregnancy test is a simple strip of paper that hides a marvel of immunoassay engineering. Unlike Cranes' test, which required that the user add reagents to the urine in a test tube, Clearblue contained all the (dried) reagents within the strip of paper. The reagents are antibodies and dyes soaked in three separate sensing lines—the reaction line, the test line, and the control line. Only the test and control lines are visible to the user. The antibodies are anchored to the lines in the visible areas and cannot move away from them. In the upstream reaction line that is not visible to the user, however, the antibodies are free to flow when urine is added: The purpose of the reaction line is to supply antibodies so they can react with the other antibodies downstream.

When a pregnant woman adds her urine to the device, she sets a sequence of reactions in motion. First, capillary action brings the hormone hCG with fluid to the hidden reaction line. The free antibodies in the reaction line bind to hCG and join the urine stream. When the stream reaches the test line, the surface-linked antibodies recognize the hCG (carrying the free-flowing antibody). Thus, the hCG molecules become immobilized in a "sandwich" between two antibodies. An enzyme dangling off one of the antibodies reacts with the dye in the test line, producing a coloration. However, not all the free-flowing antibodies bound to hCG react. As the urine keeps advancing, it carries the excess hCG-bound antibodies, which arrive at the control line and perform the same reaction as in the test line. That's a two-line-means-yes result (Figure 12.3).

If the woman had not been pregnant instead, the antibodies against hCG in the reaction line would not have been able to bind to hCG. The antibodies would have flowed past the test line without binding (or reacting with the enzyme). Still, they would have ended up, all the same, being captured by the antibodies

---

[12] FolgerLibrary. Syrup of violets and science. *YouTube.* https://www.youtube.com/watch?v=pdEbMBe0aa8 (2011).

[13] Maumené, E. J. Sur un nouveau réactif pour distinguer la présence du sucre dans certains liquides. *J. Pharm.* 17: 368–370 (1850).

[14] Oliver, G. On bedside urinary tests: Detection of sugar in the urine by means of test papers. *Lancet* 121: 858–860 (1883).

[15] Gutzeit, G. Sur une methode d'analyse qualitative rapide; Methodes d'analyse rapide à la touche des cations et anions les plus usuels. *Helv. Chim. Acta* 12: 829–850 (1929).

[16] Feigl, F. *Qualitative Analyse mit Hilfe von Tüpfelreaktionen.* Akademische Verlagsgesellschaft, 1931.

[17] Clegg, D. L. Paper chromatography. *Anal. Chem.* 22: 48–59 (1950).

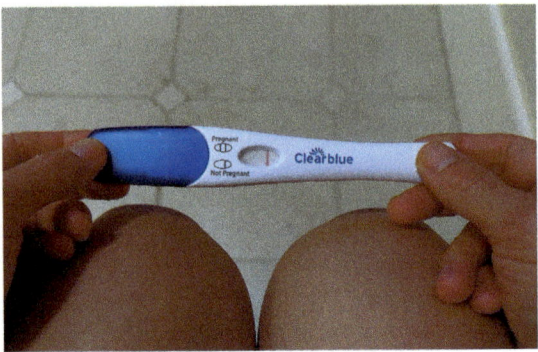

Figure 12.3 A pregnancy test.

in the control line, where the enzyme would have reacted with dye and triggered a coloring reaction. That would have been the one-line-means-no result.

Clearblue makes immunoassays much more straightforward by reducing them to a microfluidic paper strip. It is self-powered and easily interpreted. It is so portable and simple that anybody can operate and read it anywhere without batteries. The pump is the paper itself, made of tiny water-loving fibers that encourage the wetting of the fluid and pores that pull the liquid at every corner (see Figure 11.3). The pregnancy test detects a human hormone with an antibody immobilized onto a strip of paper inside the test.[g] The specificity of the test to detect pregnancy relies on the antibody against hCG, so this type of microfluidic assay can be tailored to detect many other conditions straightforwardly by changing the antibody.

We were recently unwilling witnesses to how the simplicity of microfluidics helped us all. During the COVID-19 pandemic, manufacturers of pregnancy tests were able to quickly start fabricating large numbers of at-home COVID-19 tests—a matter of international urgency to fight the virus—simply by swapping the hCG antibody in a pregnancy test with an antibody against a protein on the surface of the COVID-19 virus (Figure 12.4, left). This test is not as accurate as the polymerase chain reaction test (which amplifies the virus DNA), but it is much faster, easier, and cheaper, and it works well when you have a lot of viruses in your system. Most critically, it can be done at home, without specialized equipment at the clinic. All you need to do is collect nasal mucus with a swab and treat it with a liquid containing salt and soap that breaks down the cells and mucus so the antibody can more easily grab onto the protein on the virus' surface. A few drops suffice to run the test.

---

[g] Modern pregnancy tests use different types of paper strips (cellulose pads, glass fiber pads, and nitrocellulose strips) layered on top of each other to optimize the different reactions.

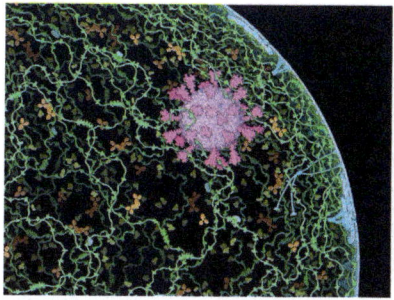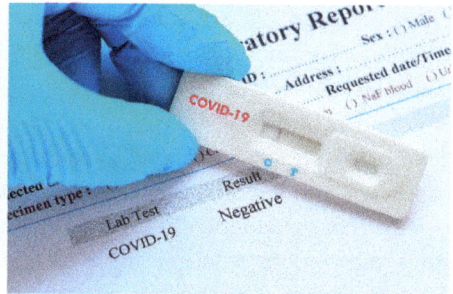

Figure 12.4 (Left) Illustration of a respiratory droplet, showing mucins (green), surfactant proteins and lipids (blue), and a COVID-19 virus (pink). (Right) A COVID-19 test, a microfluidic device similar to the pregnancy test.

Image sources: (Left) User David Goodsell on Wikipedia. https://en.wikipedia.org/wiki/ Respiratory_droplet#/media/File:Respiratory_Droplet_with_SARS-CoV-2.jpg (CC BY 4.0). (Right) Jarun Ontakrai/Shutterstock. https://www.shutterstock.com/image-photo/negative-test- result-by-using-rapid-1656883729.

Many of us have watched the fluid advancing along the strip, lighting up the control "C" line, and wishing the test "T" line would not light up next. Capillary action powers the device and draws the fluid through the network of fibers in the paper strip (see Figure 11.3). You don't need batteries. More than 2 billion easy-to-use tests (Figure 12.4, right) were distributed worldwide during the pandemic to help people monitor their exposure to the virus. Without the help of microfluidics, it would have been very difficult to overcome that nightmare.

* * *

Running cheap diagnostic assays made of paper—inspired by Clearblue—had been on the mind of George Whitesides, an eminent professor of chemistry at Harvard University, for quite some time. However, the project would not materialize until PhD student Andres Martinez joined his group in 2004. Martinez was born in Oakland, California. His Bolivian grandparents had sent his father to California in the 1970s to shield him from the brutal political repression in Bolivia at the time. Martinez's parents met in California and joined the hippie movement in Berkeley. When Andrés was five years old, the family returned to Bolivia so the four children could get a better life and education in the countryside. "I grew up raising chickens and harvesting fruits and vegetables," Martinez recalls.[18]

---

[18] Folch, A. *Hidden in Plain Sight: The History, Science, and Engineering of Microfluidic Technology.* MIT Press, 2022, 228, 231.

But Martinez did not want to be a farmer. At eighteen years of age, he left home for a community college east of San Francisco, from which he transferred to Stanford University to graduate in chemistry. He was then accepted into Harvard for a PhD in 2004. During the open-house activities offered to all incoming students, Whitesides presented a project he called "Simple Solutions" that struck a chord in the soul of farm-raised Martinez: the idea of developing diagnostic systems that (as eloquently put by Whitesides in a television interview of the time) "can work in economically constrained environments—systems that have to be affordable, workable in a low-resource setting, and must produce actionable information."[19] Martinez instantly thought, "This lab is for me."[18] Whitesides offered him a PhD slot on the spot.

Whitesides had been championing microfluidic devices for a few years but was worried they could not be used to run diagnostic tests in rural settings and remote villages of developing countries. As a chemist, he liked the operational simplicity of the pregnancy test. Nothing is simpler and easier to use than a paper strip that pumps fluid by capillarity. However, this test is designed to provide a qualitative (yes/no) answer and only takes a single fluid input, so it cannot mix reagents. For many diagnostics, doctors want to run a quantitative assay that indicates how sick the patient is. Such complex assays—involving mixtures and timing of reagents—demand the compartmentalization of fluids in separate volumes. By contrast, the pregnancy test is based on only one rectangular strip—a single fluidic compartment of limited functionality. Whitesides proposed to Martinez to work on techniques to compartmentalize fluids on paper and gave him ample leeway to achieve that goal.

For a few months, Martinez tried depositing water-repellent solutions using pens and printers but had been unsuccessful. When Whitesides saw the results, he suggested trying more established high-resolution techniques such as photolithography, the method used to pattern microelectronic chips (see Figure 2.4). Photolithography involves patterning a thin, photosensitive layer of *photoresist* by exposure to ultraviolet light. There was one problem. They were thinking of using chromatography paper, a type of paper made of pure cellulose that chemists have used for a long time to filter and separate solutions. However, this paper was not allowed in the cleanroom facility at Harvard because it sheds particle residues: Paper is not considered "clean." Martinez decided that the only way to run his experiment would be to smuggle the paper into the facility at 5 a.m. when no one else was working. After all, one cannot simultaneously start a revolution and be a rule-follower. (Martinez, now a respected professor but still

---

[19] CGTN. Simple solutions (George Whitesides Interview). *YouTube.* https://www.youtube.com/watch?v=fd21o-oaENY (2017).

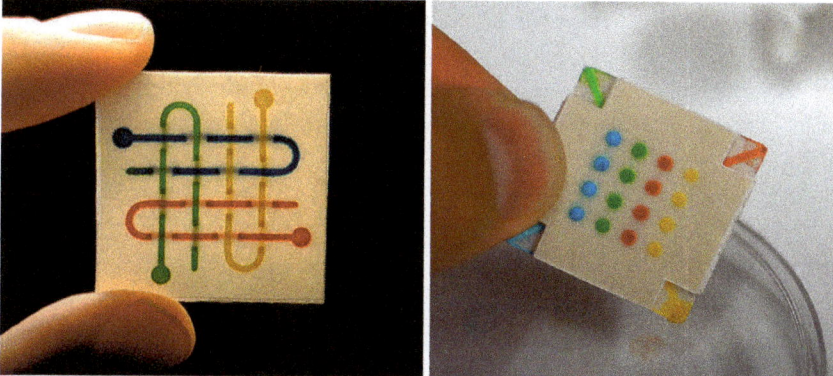

Figure 12.5 Multilayered paper microfluidic devices. (Left) A braided 3D network of four non-intersecting microchannels in paper. (Right) Each color dye is loaded in a different corner of the square and connects to four dots through a 3D network of paper-based microchannels.

Images courtesy of Andres W. Martinez, California Polytechnic State University.

looking like the young, enthusiastic student willing to take a chance on a novel paper microfluidics project when no one else would, laughed when I asked if anybody at Harvard had questioned how he had managed to pattern paper in a cleanroom: "Nobody noticed!"[18]) Once inside the cleanroom, he poured the photoresist on his chromatography paper disk and allowed the photoresist to soak into the paper. Then he developed it and observed the photoresist pattern magically appear under the yellow light of the cleanroom—a photoresist pattern had become embedded inside the paper, rather than forming atop the surface.

The voids left by the photoresist pattern inside the paper effectively created "paper channels" that he filled with a colored dye. Making different patterns proved so easy that he made several overlapping channels in a 3D crisscrossing design (Figure 12.5). He built a device that measured the presence of glucose using only 5 microliters—approximately the volume of a teardrop—of urine.[20] As expected, capillary action filled the channels and chambers where the biochemical reactions occurred spontaneously—no batteries required.

The new technique was a huge success. Soon, several other group members rushed to expand the technique's capabilities, and a large number of researchers started using these methods precisely because they were so inexpensive and simple to implement—just as envisioned by Whitesides. A few years after Martinez

[20] Martinez, A. W., Phillips, S. T., Butte, M. J., & Whitesides, G. M. Patterned paper as a platform for inexpensive, low-volume, portable bioassays. *Angew. Chemie-Int. Ed.* 46: 1318–1320 (2007).

finished his PhD in 2010, paper-based biochemical assays to diagnose malaria,[21] tuberculosis,[22,23] and influenza[24] in resource-poor settings were no longer a dream. Many companies now distribute the technology. Much like Murugananntham, Martinez, the scientist who once was a farm boy, revolutionized global health with microfluidics.

Summary

- The wicking properties of cellulose have been used to power many sanitary products, especially in women's health.
- Upon learning about his wife's period for the first time, Arunachalam Muruganantham, from a small town in rural India, became determined to help her. Without having ever set foot in a university, he ended up learning how to manufacture hygiene pads at low cost and understanding why some materials wick better than others. Muruganantham's award-winning machines are now used all over Indian rural communities to fabricate hygiene pads with cellulose from tree pulp.
- The first portable pregnancy test was developed by Unilever in 1988 and consists of an immunoassay for the presence of human chorionic gonadotropin (hCG) in the woman's urine. During the assay, flow is powered by the paper strip itself—in other words, by capillary wicking. The rapid test for COVID-19 works with the same principles.
- In the 1990s, inspired by the hCG immunoassay, George Whitesides and his graduate student, Andres Martinez, at Harvard University developed *paper microfluidics* to run inexpensive diagnostics, an approach that has revolutionized global health.

[21] Reboud, J., Xu, G., Garrett, A., Adriko, M., Yang, Z., Tukahebwa, E. M., et al. Paper-based microfluidics for DNA diagnostics of malaria in low resource underserved rural communities. *Proc. Natl. Acad. Sci. USA* 116: 4834–4842 (2019).
[22] Veigas, B., Jacob, J. M., Costa, M. N., Santos, D. S., Viveiros, M., Inácio, J., et al. Gold on paper— Paper platform for Au-nanoprobe TB detection. *Lab Chip* 12: 4802–4808 (2012).
[23] Shetty, P., Ghosh, D., Singh, M., Tripathi, A., & Paul, D. Rapid amplification of *Mycobacterium tuberculosis* DNA on a paper substrate. *RSC Adv.* 6: 56205–56212 (2016).
[24] Anderson, C. E., Buser, J. R., Fleming, A. M., Strauch, E. M., Ladd, P. D., Englund, J., et al. An integrated device for the rapid and sensitive detection of the influenza hemagglutinin. *Lab Chip* 19: 885–896 (2019).

PART III

GOING WITH THE FLOW

Enthusiasm is the engine of success.

—Ralph Waldo Emerson (1803–1882)

We have seen how plants photosynthesize sugars on the spot and use capillary action to pump nutrients to every leaf—they are autonomous, at least by comparison to animals. Animals, on the other hand, need to constantly forage for food and water to survive. This continuous motion depends on intense energy-producing reactions (collectively called *metabolism*) that require providing nutrients and energy to all the organism's cells at a fast pace—a task for which capillarity, a slow process, is an underperformer. For this purpose, large animals have evolved a specialty organ (the heart) that helps them pump the nutrient liquid (the blood), harvest oxygen from the environment (the gills or lungs), synthesize essential biomolecules (the liver), and filter out the chemical waste generated as a by-product of metabolism (the kidneys). All these organs continuously process large amounts of fluid flow (usually blood) through many tiny channels in parallel, collectively known as *vasculature*, the smallest of these being the *capillaries*. Microfluidic engineers have used this paradigm as inspiration to build devices for improving the mixing efficiency of chemicals, emulating drug testing outside the organism, and personalizing better disease treatments.

At small scales, fluids behave unintuitively—unlike the topsy-turvy turbulent flow gushing down a waterfall, twisting out of a garden hose, or splashing into our bathroom or kitchen sinks that we experience almost daily at larger scales (Figure PIII.1). Instead, flows are eerily stable in the constricted volumes of a submillimeter-sized conduit, such as a blood capillary. In a microchannel, the walls are so close to the center that they "hug" the streamlines into organized parallel streams, called *laminar flow* (Figure PIII.2). This marvelous stability appears by a spirited interplay between inertia and viscosity.

Inertia has been well understood since 1687 when Isaac Newton formulated the laws of motion in his *Principia Mathematica*. Inertia, Newton stated, is the tendency of objects to keep their state of motion in the absence of forces. This

Figure PIII.1 We are very familiar with the turbulent flows encountered at large, everyday scales. In turbulent flow, the paths followed by fluids are chaotic, which means they are mathematically unpredictable.

Image source: Pxhere.com. https://pxhere.com/en/photo/878077.

concept is easy to grasp with a cannonball or a baseball, but it gets a bit trickier when you try to hold a fluid, which deforms and slips between your fingers. And we all know we cannot throw a rock under water. Newton was well aware of these facts and even provided the first working formula for viscosity, the property of a liquid that describes its resistance to flow. Newton famously formulated the second law of motion, which stated that the force (F) applied to an object of mass M will give it an acceleration equal to a such that $F = m \times a$. This formula successfully predicted a wide range of phenomena, from the parabolic trajectory of a thrown ball to the elliptical orbit of a planet around its star. However, many generations of scientists after Newton were baffled by a perplexing mathematical conundrum: Newton's second law of motion was not straightforwardly applicable to liquids.

It took French engineer and physicist Claude-Louis Navier and Anglo-Irish physicist and mathematician George Gabriel Stokes several decades between the 1820s and 1850 to develop the necessary mathematical tools. They adapted Newton's $F = m \times a$ into a set of equations that describe the movement of a viscous fluid such as water. These equations are called the *Navier–Stokes equations* in their honor. It's not a perfect tool yet, but it works well in many cases—and it's all we have now.

Figure PIII.2 Laminar flow can be observed at all scales, but it is particularly common—and awe-inspiring—in microfluidic systems such as these 100-micrometer-wide channels. Fluids flow from bottom to top. Fluids and particles in laminar flow follow paths that are mathematically predictable—a necessity for precision engineering and medical device design.

Image source: Greg Cooksey and Albert Folch.

One of the more remarkable predictions of the Navier–Stokes equations is that for small pipes and small pressures, all the flow lines go parallel—the flow is "laminar," as commonly observed in microchannels (see Figure PIII.2). That's all the magic that keeps the streams from mixing. In this laminar flow regime, the frictional forces due to viscosity far exceed the "inertial" forces that cause the fluid to move—we witness the liquid's most sluggish behavior. For example, flow in blood vessels is mainly laminar. When zooplankton swim in the ocean, they beat their flagella to displace water in a laminar mode, similar to how a human would struggle to swim through mud or molasses. At a microscopic scale, viscosity and laminar flow dominate the fluid world.

At a large or macroscopic scale, on the contrary, the inertial forces tend to overcome the viscous forces and disrupt the nicely layered organization of the fluid. We see a rapidly changing, turbulent flow. A giant whale swallowing those same zooplankton uses this turbulence to its advantage: It swims and turns through water aided by the eddies and swirls generated by its tail and fins.[1] For the whale, the ratio of the viscous forces it feels compared to the inertial forces it generates is 10 billion times smaller than for the zooplankton.[2] Unfortunately, the Navier–Stokes equations fail miserably when we try to predict the shapes of breaking waves and waterfall splashes. Befuddlingly, Nature seems to punish us by making it difficult to predict the drag when we swim in water, among many daily observations. Fluid mechanics still cannot explain why the drag of a simple cylinder decreases dramatically in a fluid at very high flow rates (the "drag crisis") or why a well-designed airship can have less than 2% of the drag of a sphere of the same frontal area ("streamlining").[2] Yet we can easily explain the flagellar swimming (also in water) of creatures thinner than our hair.

Perhaps less obvious, but more intimately ours, is that all complex multicellular organisms owe their very existence to microfluidics. During development, they all rely on microscopic gradients of molecules called *morphogens* to attain their final, convoluted shape. Secreted by nearby cells, the diffusion[a] of these morphogens tells the other growing cells where they are, in which direction they should grow, and where and when they need to stop growing.[3] The body grows according to a predetermined microfluidic canvas of morphogens that changes by the minute. One of the most intricate structures in this body architecture is the vasculature. Look at any part of your body, or of any multicellular creature, with a magnifying lens, and you will see that a hair, a root, a stem, a limb, an organ could not have formed without complex networks of tiny distribution channels that bring nutrients and signaling molecules to every cell of the organism through the sap or blood (Figure PIII.3). And it's not just the blood—in

---

[a] Other mechanisms, such as the degradation of morphogens and their active transport by nearby cells, can also play a role in the formation of a gradient.[3]

[1] Fish, F. E., & Lauder, G. Not just going with the flow. *Am. Sci.* 101: 114 (2013).

[2] Vogel, S. *Life in Moving Fluids: The Physical Biology of Flow.* Princeton University Press, 1994.

[3] Wartlick, O., Kicheva, A., & González-Gaitán, M. Morphogen gradient formation. *Cold Spring Harb. Perspect. Biol.* 1: a001255 (2009).

Figure PIII.3 The face vasculature of a human cadaver can be visualized by filling the veins and arteries with a plastic resin, a technique called plastination.

Image source: Wallpaperflare.com. https://www.wallpaperflare.com/germany-ulm-blautalcenter-veins-body-worlds-koerperwelten-wallpaper-evrye/download.

organs such as the liver and the kidney, there are parallel webs of tiny conduits to usher waste fluids such as urine and bile out of our body.

We *are* microfluidic.

13

Dirty Blood

The Amazing Microfluidics of the Kidney, Kidney Dialysis, Hypodermic Needles, and Mosquitos

David Rush, a Black rapper, successful recording artist, and music producer from New Jersey, started having problems with his kidneys in high school. It was not until later, when he was twenty-four years old and was scheduled to go on tour with his good friend and famous rapper Armando Christian Pérez (also known as Pitbull), that he realized he needed to be more proactive about his health. Truth be told, he had been hiding his condition for a while. In tenth grade, he was not allowed to play football after doctors spotted blood and protein in his urine—indications of chronic kidney disease. However, he was allowed to play in his junior and senior years and felt fine. He then went to college and did the usual things that college students do: partying, drinking, some studying. After that, he started working. Just as his career was beginning to take off, a routine medical exam revealed that his kidneys had been silently deteriorating to the point of no return.[1]

"My doctor told me that if I didn't start this thing called *dialysis*, I wouldn't live past a year," said Rush.[2] Dialysis is a treatment that artificially replicates some of the work done by the kidneys by diverting the blood flow through a machine, which operates on microfluidic principles. "I was scared."[2] His doctor told him he had chronic kidney failure and he only had about one year to live if he didn't act quickly on it. He started going for dialysis sessions at the hospital—three times a week, a horrendous five hours each session. "That was the equivalent of a part-time job," he wrote later.[3] He also did not want to share his condition with others. "I thought that if I told anyone, I would be seen as weak, so I would schedule my work around my dialysis and kept it a secret," he said.[3]

Pitbull's offer to go on tour rushed things. First, he told Pitbull there was no way he could make the tour, but to his surprise, Pitbull insisted. Rush decided to

[1] Walden, T. Rapper David Rush talks battling kidney failure & being forever thankful for his big brother. BlackDoctor.org. https://blackdoctor.org/rapper-david-rush-talks-battling-kidney-failure-being-forever-thankful-for-his-big-brother (2016).
[2] NxStage Medical. David's journey with NxStage HHD. YouTube. https://www.youtube.com/watch?v=dFiwcrXPTWY.
[3] Rush, D. The raw truth from a home hemodialysis warrior. *Outset*. https://www.outsetmedical.com/perspective/the-raw-truth-from-a-home-hemodialysis-warrior (2021).

come clean about his diagnosis and take with him a portable dialysis machine. However, the portable device required substantial discipline to use correctly.[4] He had to store fifty boxes of supplies in his house, filling a whole room. Before treatment, five days a week, he would have to prepare a six-hour batch of *dialysate*, the fluid that helped clean Rush's blood in the dialysis machine. It was overwhelming, so in early 2020, he decided to go back to the hospital for dialysis treatment. It turns out that both his kidney and the dialysis machine keeping him alive utilize microfluidics.

The kidneys are very smart microfluidic sieves. They primarily act like sentinels, regulating the amount of water and salts in and out of the blood. Salts comprise various essential chemicals, electrically charged dissolved minerals (*electrolytes*), and not just the ordinary table salt (*sodium chloride*) in your kitchen shaker. These electrolytes—sodium, potassium, calcium, magnesium, chloride, and bicarbonate—are essential for nerve and muscle function and for balancing the right amount of acid and base in the blood and other body fluids.[5] The water adjustment is critical to maintain a constant blood volume and pressure.[6] Due, in part, to our healthy kidneys, we do not get bloated every time we drink a big mug of beer or glass of water. The kidney's filter also cleans out waste that flows through our body: It removes the toxins that we ingest daily or that are produced by bacteria inside our bodies when we have an infection and excretes them into our urine along with any excess water. All these processes happen in—and thanks to—the microfluidic channels of the kidney.

Each kidney is a bean-shaped organ the size of a fist with three big fluidic connections: one input and two outputs. The input is the large *renal artery* that takes in the "dirty" unfiltered (but freshly oxygenated) blood.[a] One output is the *renal vein* returning the clean, filtered blood to the body. The other output is the *ureter*, which acts as the body's sewage drain and delivers urine drop by drop to the bladder. In twenty-four hours, a pair of healthy kidneys filter 200 liters of blood—the equivalent of a full bathtub—and remove approximately 2 liters of water, toxins, and other waste as urine. That is a lot of uninterrupted day-and-night filtering for a couple of avocado-size organs.

How can the kidneys filter and process so much blood? The key is massively parallel processing by microfluidics. Inside each kidney, each renal artery fans out like a tree into roughly half a million tiny capillaries. In most vertebrates,

---

[a] As it is pumped out of the heart, oxygenated blood rushes down the center of our abdominal cavity through the aorta, a 3-centimeter-wide artery from which the two 5-millimeter-wide renal arteries separate, taking approximately one-third of the total blood flow. Hence, the kidneys only clean up part of the blood volume at a time.

[4] Ellis, H. Rapper David Rush's patient perspective on kidney disease. Giddy. https://getmegiddy.com/david-rush-kidney-disease (2022, May 20).

[5] Weiner, I. D., & Verlander, J. W. Ammonia transporters and their role in acid–base balance. *Physiol. Rev.* 97: 465–494 (2017).

[6] Scott, R. P., & Quaggin, S. E. The cell biology of renal filtration. *J. Cell Biol.* 209: 199–210 (2015).

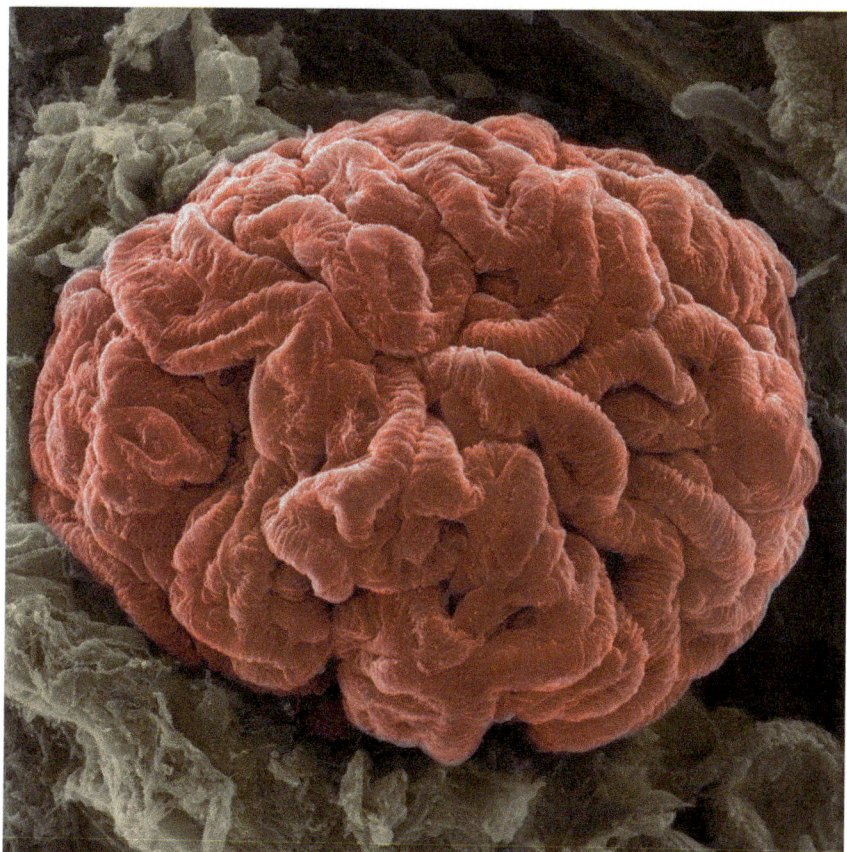

Figure 13.1 A kidney glomerulus from a rhesus monkey. This ball-of-yarn tangle is formed of a single capillary blood vessel (false-colored in red) close to one hundredth of a millimeter in width. The blood rushes in full of urine and inside this tangle the urine drips out, so the blood leaves out the glomerulus clean of urine. The capsule that collects the urine is partially visible (false-colored in brown).

Image source: www.HistologyGuide.com, T. Clark Brelje and Robert L. Sorenson, University of Minnesota, Minneapolis, MN.

each capillary ends in a *glomerulus* (Figure 13.1), a microscopic ball-of-yarn tangle surrounded by a capsule that collects the filtrate. Glomeruli did not start existing all of a sudden—early in evolution, most animals had rudimentary proto-kidneys that served their filtering purpose; Nature tinkered with those by trial-and-error in stages over millions of years until they evolved into the present kidneys,[7] a bit like the microfluidics of plants (see Chapter 7). As an exception,

---

[7] Natochin, Y. V. Evolutionary aspects of renal function. *Kidney Int.* 49: 1539–1542 (1996).

some marine bony fishes such as the seahorse have kidneys that do not have glomeruli, but amphibians, reptiles, birds, and mammals all have glomeruli.[8] This microfluidic marvel, of which humans have roughly 1 million duplicates between our two kidneys, is where most of the action happens.

Glomeruli owe their efficiency to their minute size. Untangled, each yarn would be almost 1 centimeter long, but the diameter of the balled-up glomerulus is only between a quarter and a tenth of a millimeter. The capillary walls forming the glomerulus are made of just one cell in thickness. These specialized cells have tiny pores that let fluid and small molecules pass through, but not large proteins or cells.

The kidneys' microfluidic body-maintenance service does not always work to perfection. These processes are so vital for maintaining our body healthy that if both of our kidneys were to cease functioning suddenly, we would die within a matter of days. Sometimes the kidneys malfunction sporadically, producing painful kidney stones. In the case of chronic kidney disease—as for David Rush—they stop functioning gradually, giving doctors time to intervene. Unfortunately, Rush's condition is more common than we might think. More than half a million people in the United States alone[9] and 2 million worldwide[10] are living with dialysis right now.[b] Although not perfect, dialysis—essentially, an artificial microfluidic blood cleansing process that substitutes the body's microfluidic natural one—allows them to stay alive.

The dialysis machine for people like Rush with dwindling or no renal function was invented by young Dutch physician Willem Kolff in 1943. While Kolff was helping fight the German Nazi occupation during World War II, he built his ingenious dialyzer prototype from the scarce odds and ends available to him at the time: sausage skins, orange juice cans, a washing machine, and other everyday items one would never dream of adding to a medical device. Over the next two years, Kolff tried but failed to save any patients with acute kidney failure with his new instrument. Finally, in 1945, after eleven hours of hemodialysis with his machine, he was able to bring a sixty-seven-year-old out of a coma caused by kidney failure. After the war ended, he donated his machines to hospitals around the world, where they became the standard of care.

---

[b] This number may only represent 10% of the people who actually need treatment to live.[10]

[8] Kimball, J. W. 15.5B: Vertebrate kidneys. LibreTexts. https://bio.libretexts.org/Bookshelves/Introductory_and_General_Biology/Biology_(Kimball)/15%3A_The_Anatomy_and_Physiology_of_Animals/15.05%3A_Excretion/15.5B%3A_Vertebrate_Kidneys (n.d.).

[9] National Institute of Diabetes and Digestive and Kidney Diseases. Kidney disease statistics for the United States. National Institutes of Health. https://www.niddk.nih.gov/health-information/health-statistics/kidney-disease (2023).

[10] National Kidney Foundation. Global facts: About kidney disease. https://www.kidney.org/kidneydisease/global-facts-about-kidney-disease (2015).

All dialysis machines since Kolff's have thin, artificial semipermeable membranes made of polymers such as cellulose (in place of sausage skin) or cellulose acetate. These thin membranes function similarly to the ones at the kidney's glomeruli to remove substances dissolved in the blood. Three processes co-occur during dialysis. First, water moves across the membrane to equalize the concentration, a process called *osmosis*. Second, the different ions also move across the membrane to equalize their concentration, a process called *diffusion*. Third, when additional pressure helps accelerate these movements, that is called *ultrafiltration*. All three processes also occur naturally in the kidney, although the kidney cells that line the ducts downstream of the glomerulus use additional, more sophisticated mechanisms to regulate electrolytes.

Researchers presented many new designs in the 1960s. Kolff's original rotating-drum, washing-machine-like prototype was mechanically complex and prone to failure. A team led by Dr. Belding Scribner in Seattle, Washington, developed a new design during a four-hour meeting around a coffee table in the back of a plane, which became the first commercial home dialysis machine. As a result of these efforts, patients could operate dialysis machines from the comfort of their homes, giving them more freedom and flexibility.[c] Home dialysis made patients feel better because they could dialyze their blood daily, with shorter treatment sessions (two or three hours), giving them more energy to do the things they loved. That demand for home dialysis required reliable, user-friendly systems.

The Scribner design eventually evolved into a "capillary kidney,"[11] now called the *hollow fiber* design,[d] with no moving parts (save for the pump) and thus less prone to failure or contamination. The first patient to be treated with a hollow fiber dialysis machine was at the Marquette School of Medicine in Milwaukee, Wisconsin, in 1967.[12] Most dialysis machines today use this microfluidic design. The patient's blood is pumped through a cylinder containing a bundle of approximately 10,000 capillaries of hollow polymer fibers bathed in fluid (Figure 13.2, top). Each fiber has an inner diameter of about 200 micrometers. Its wall, a thin porous membrane approximately 20 micrometers thick, acts like a filter (Figure 13.2, bottom).[13] The blood runs through the fibers, and the toxins,

---

[c] In-hospital dialysis is typically based on three- to four-hour-visits three times a week, leaving patients tired on the days off.

[d] This design is also known as the *Dow dyalizer* because Dow Chemical Company, which in the 1960s already had projects to develop hollow fiber membranes for various applications, was selected as the technical contractor to develop it commercially.

[11] Stewart, R. D., Lipps, B. J., Baretta, E. D., Piering, W. R., Roth, D. A., & Sargent, J. A. Short-term hemodialysis with the capillary kidney. *Trans. Am. Soc. Artif. Intern. Organs* 14: 121–125 (1968).

[12] Sargent, J. A. Dialysis in the 1960s and the first hollow fiber dialyzer. *Int. J. Artif. Organs* 30: 953–963 (2007).

[13] Sakai, K. Dialysis membranes for blood purification. *Front. Med. Biol. Eng.* 10: 117–129 (2000).

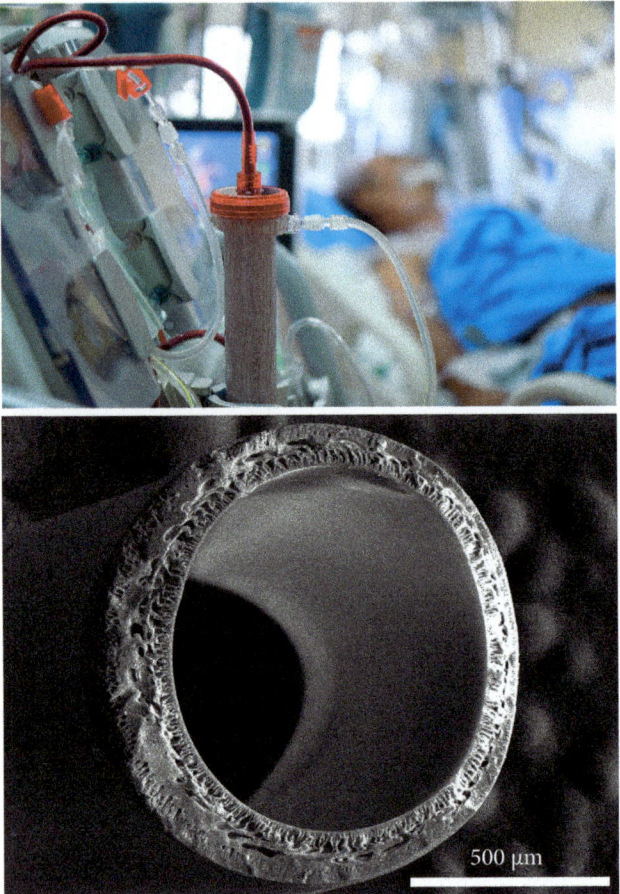

Figure 13.2 (Top) A hollow fiber dialysis machine. (Bottom) Close-up of a single hollow fiber of nearly half a millimeter in diameter.

Image sources: (Top) Hospital man/Shutterstock. https://www.shutterstock.com/image-photo/ dialysis-machine-working-acting-substitute-kidneys-1960968457. (Bottom) User RobertsBiology on Wikipedia. https://en.m.wikipedia.org/wiki/File:Polysulfone_hollow_fiber_membrane_SEM_cross-section.png (CC BY-SA 4.0).

waste, and excess water seep through the pores of the walls into the outside fluid. This "dirty" fluid—like urine—is removed from the cylinder through a side port. With massively parallel microfluidic filters, this device acts as an artificial kidney.

More recently, Rush has been lucky to benefit from the latest, user-friendly generation of portable dialysis machines, with minimal supplies and no need to batch fluids. To this day, he can live an otherwise everyday life. Like him, all dialysis patients rely on this microfluidic technology to clean their blood daily. "We would wake up early to travel to our next city," Rush recalls. "I would set myself up in my hotel for a four-hour treatment before packing everything up. I would then do an hour set, opening for Pitbull. He would do his set, and we would be back on the road."[4] Rush, a caring father and husband, did not miss treatment or a show and accepted his disability. He even articulated it with his rap music to become an advocate for others with the same condition:

> Home hemo is the way to go
> Educate y'rself and get into know
> Don't let your therapy j'st steal your show
> Slow down the flow and give back the glow
> You could do treatment inside of your own home.[14]

* * *

None of us, if we had a choice, would want to be stuck by a needle, so imagine what people like David Rush or diabetics (see Chapter 14) have to go through every day to sample their blood. Doctors have only been using needles since the 1850s, but they have been using them a lot—an estimated 16 billion procedures since then.[15] Although the idea of injecting people had been tried before, Irish physician Francis Rynd is credited with the first successful subcutaneous—later termed *hypodermic*—injection in 1844 on Ms. Margaret Cox in the Meath Hospital in Dublin, Ireland. The outer diameter of hypodermic needles (Figure 13.3)—also known as the *Birmingham gauge* or, simply, gauge[e]—has been gradually reduced to minimize pain and bleeding.[15] Nowadays, children are vaccinated with 25-gauge needles that measure just about half a millimeter in diameter (514-micrometer outer diameter and 260-micrometer inner diameter). Finer needles of 34 gauge with 184-micrometer outer diameter and 83-micrometer inner diameter also exist. These truly microfluidic needles hurt much less.

Although human engineering skills deserve a lot of praise, Nature has had millions of years to try to evolve solutions that are often better. Mosquitoes bite us and extract blood from us without us even noticing. How do they do it? Although a mosquito bite can be a health concern, it is one of the most amazing microfluidic marvels of the animal kingdom. Whereas male mosquitos are essentially vegetarian, the female mosquito drills into your skin with her

---

[e] Originally developed in early nineteenth-century England for the manufacture of metal wires, in medicine the Birmingham gauge specifies the outer diameter of hypodermic needles, catheters, cannulae, and suture wires. The gauge number increases with decreasing diameter and runs from zero (corresponding to 12.7 millimeters) to thirty-six (corresponding to 0.1 millimeters).

[14] NxStage Medical. Home hemodialysis rap with David Rush. YouTube. https://www.youtube.com/watch?v=gikJXGhmc-w (n.d.).

[15] Gill, H. S., & Prausnitz, M. R. Does needle size matter? Hypodermic needles. *J. Diabetes Sci. Technol.* 1: 725–729 (2007).

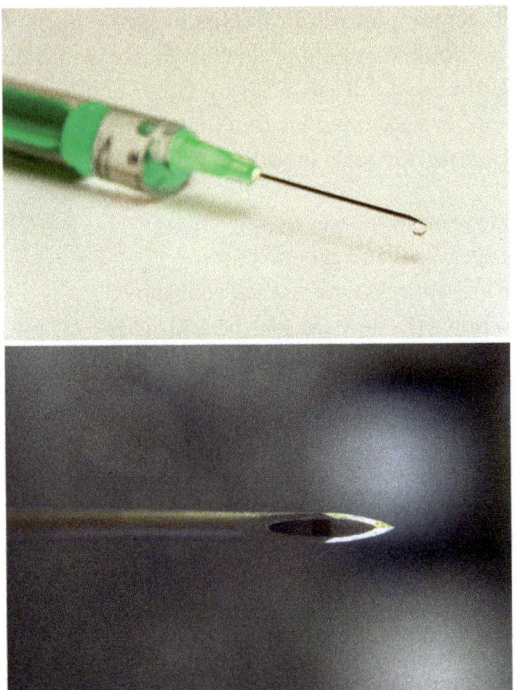

Figure 13.3 (Top) Hypodermic needle mounted on a syringe. (Bottom) Close-up of the tip of a hypodermic needle.

Image sources: (Top) Pxhere.com. https://pxhere.com/en/photo/707717. (Bottom) User Yarzaryeni on Wikipedia. https://en.wikipedia.org/wiki/Hypodermic_needle#/media/ File:Hypodermic_needle_tip.png.

2-millimeter-long *proboscis*, a sort of ensheathed flexible sword that acts as a mouthpart.[f] The sheath or *labium* acts as the mosquito's "lips" and ends in the *labella*, a structure with sensors that detect the most suitable skin spot to dig for blood. When the labella identifies the ideal location, the labium retracts, and the sword or *fascicle* starts drilling (Figure 13.4, top). The fascicle is a marvelously complex microscopic needle, unparalleled by human ingenuity, three times thinner than one of your hairs, and made of chitin protein (Figure 13.4, bottom).

The fascicle seems to have been conjured up by a Hollywood special effects team for a nanorobot microsurgery movie.[16] It comprises six movable stylets. The two *maxillae* end in tiny serrated teeth that help the mosquito saw through the skin like a steak knife. Another pair, the *mandibles*, hold the tissue apart

---

[f] Less protein-hungry male mosquitos feed on the safer nectar juices of plants and will not bite humans because their less sophisticated proboscis cannot penetrate skin.

[16] Deep_Look. How mosquitoes use six needles to suck your blood. YouTube. https://www. youtube.com/watch?v=rD8SmacBUcU (2016).

Figure 13.4 (Top) A mosquito drills into the skin with her fascicle, a natural microneedle, to extract blood. As she pumps, she keeps the blood cells and filters out the liquid through the bottom of her abdomen. (Bottom) Scanning electron micrograph of the fascicle of a yellow fever mosquito. This fascicle and the hole are approximately 20 and 12 micrometers wide, respectively.[19]

Image sources: (Top) Pxhere.com. https://pxhere.com/en/photo/921517. (Bottom) Courtesy of Dr. Melur (Ram) K. Ramasubramanian, North Carolina State University.

like chopsticks while she goes at it. The last two stylets are hollow tubes and function like straws. She uses the most sophisticated hollow one, the *labrum*, to sniff around the chopped tissue until she finds a blood vessel. She uses the other hollow stylet (*hypopharynx*, the "tongue" of the mosquito) to inject saliva into the wound with a numbing chemical. This camouflage measure allows her to proceed undetected during the drilling.[17] When the tip of the labrum

---

[17] Mosquito Magnet. How mosquitoes bite. https://www.mosquitomagnet.com/articles/how-mosquitoes-bite (n.d.).

finds a blood vessel, it pierces the vessel, and she starts pumping blood into her abdomen with it. The saliva delivered by the hypopharynx also contains an anticoagulant to prevent blood clot formation.[g] It is this anticoagulant, not the drilling, that causes inflammation and pain.[18] As she pumps, she discards the water from the blood through her back (Figure 13.4, top) and only keeps the protein-rich blood cells—a two-minute meal that will feed her eggs.[16]

Unfortunately, these delicate microsurgery and microfluidic suction operations that the female mosquito devotes to nurturing her progeny have enormous, unintended consequences for others. Unbeknownst to her, on certain occasions, the blood she pumps into her abdomen from her host picks up tiny hitchhiker germs—bacteria, viruses, or parasites. The germs happily fly away with her and live inside her for a while, eventually making their way into her saliva. The next time the mosquito goes bloodhunting, she inadvertently injects the germs into another animal or person. Mosquitos are the most deadly animals as far as humans are concerned. Each year, mosquitoes cause illnesses in nearly 700 million people. Of these, 725,000 humans die yearly, the vast majority young children in Africa from malaria or yellow fever transmitted by mosquito bites. For all the excellent work microfluidics can do, it too has its dark side, even in Nature.

Microneedle engineers have spent long hours studying mosquitoes and emulating their microneedle designs.[19,20] Painless *transdermal microneedles* come to the rescue of patients who need frequent punctures or are afraid of needles.[h] They sometimes look like something from a science-fiction movie (Figure 13.5). One of the finest examples of micromachining technology, they can be just a fraction of a millimeter in length—enough to pierce the outer layer of dead cells

---

[g] Your immune system's reaction to the mosquito's saliva produces an itchy red spot, the characteristic mosquito bite pimple.

[h] Microfabricated needles were pioneered in the late 1990s by Mark Prausnitz and Mark Allen at Georgia Tech using silicon etching techniques borrowed from microelectronics, and they have evolved a lot since then.[21,22] Even earlier, Kensall Wise and Jim Angell (see Chapter 6) had fabricated solid needles with electrodes for neural recordings in 1970.[23] Wise's microneedles could be fabricated in arrays, but they had to be fabricated on a flat substrate so the arrays looked like a linear comb. After Prausnitz and Allen, the first out-of-plane hollow microfabricated needles required more advanced etching techniques and were achieved in 2005 by Dorian Liepmann's team at the University of California, Berkeley.[24]

[18] Dumé, B. Painless needle mimics a mosquito's bite. *New Scientist* https://www.newscientist.com/article/dn14348-painless-needle-mimics-a-mosquitos-bite (2008, July 17).

[19] Ramasubramanian, M. K., Barham, O. M., & Swaminathan, V. Mechanics of a mosquito bite with applications to microneedle design. *Bioinspiration and Biomimetics* 3: 046001 (2008).

[20] Zhou, Y., Yang, H., Wang, X., Yang, H., Sun, K., Zhou, Z., et al. A mosquito mouthpart-like bionic neural probe. *Microsystems Nanoeng.* 9: 88 (2023).

[21] Henry, S., McAllister, D. V., Allen, M. G., & Prausnitz, M. R. Microfabricated microneedles: A novel approach to transdermal drug delivery. *J. Pharm. Sci.* 87: 922–925 (1998).

[22] McAllister, D. V, Allen, M. G., & Prausnitz, M. R. Microfabricated microneedles for gene and drug delivery. *Annu. Rev. Biomed. Eng.* 2: 289–313 (2000).

[23] Wise, K. D., Angell, J. B., & Starr, A. An integrated-circuit approach to extracellular microelectrodes. *IEEE Trans Biomed Eng* 17: 238–247 (1970).

[24] Sivamani, R. K., Stoeber, B., Wu, G. C., Zhai, H., Liepmann, D., & Maibach, H. Clinical microneedle injection of methyl nicotinate: Stratum corneum penetration. *Ski. Res. Technol.* 11: 152–156 (2005).

Figure 13.5 (Left) A hypodermic microneedle entirely etched in silicon. The microneedle is about a quarter of a millimeter across and tall. The hole is approximately 25 micrometers (thousandths of a millimeter) in its narrowest dimension. (Right) A microarray of microneedles, spaced half a millimeter apart from each other.

Images courtesy of Han Gardeniers and Albert van der Berg, University of Twente, the Netherlands.

on your skin but not enough to reach nerve endings or blood capillaries. You can't even feel them, so they do not need to add a numbing solution like the mosquito. You can barely *see* them, so needle phobics do not have anything to be afraid of. Although these microneedles are less effective than the scary metallic needles, they are already used to deliver drugs and vaccines through your skin.[25] Longer microneedles can be used to draw blood. Mimicking the mechanisms of the mosquito bite, microfluidic engineers have designed microneedles that vibrate upon insertion to reduce tissue displacement for the extraction of a biopsy.[26]

Microneedles are more hygienic, easier and safer to use, transport, and dispose of (no biohazard sharpies container required). Therefore, they are more cost-effective in the long term. To make up for the tiny size, each patch typically contains many microneedles inserted at once (Figure 13.5, right). Microfluidic design is very flexible, although not quite as adaptable and efficient as that of mosquitoes, which, in all fairness, have had a head start of more than 100 million years of evolution over us.[27] Perhaps not too far in the future, nurses will use pain-free microneedles when they need to get a blood sample, and people on dialysis, such as David Rush, will have one less thing to worry about every time they prick their skin.

[25] Kim, Y.-C., Park, J.-H., & Prausnitz, M. R. Microneedles for drug and vaccine delivery. *Adv. Drug Deliv. Rev.* 64: 1547–1568 (2012).

[26] Li, A. D. R., Putra, K. B., Chen, L., Montgomery, J. S., & Shih, A. Mosquito proboscis-inspired needle insertion to reduce tissue deformation and organ displacement. *Sci. Rep.* 10: 12248 (2020).

[27] Poinar, G., Zavortink, T. J., & Brown, A. *Priscoculex burmanicus* n. gen. et sp. (Diptera: Culicidae: Anophelinae) from mid-Cretaceous Myanmar amber. *Hist. Biol.* 32: 1157–1162 (2020).

Summary

- The kidneys are microfluidic sieves that remove waste and toxic substances from the blood while regulating the body's water and salt balance.
- The glomerulus, made of special blood capillaries, is the basic filtration unit of the kidney (although there is also additional processing occurring in other parts of the kidney). Each kidney contains close to half a million glomeruli.
- More than 2 million people are on kidney dialysis worldwide as a consequence of their kidney disease.
- Kidney dialysis is based on the filtration of blood through tens of thousands of parallel microfluidic hollow fibers. Because the walls of the fibers are semipermeable, the blood can be "cleansed" in a way that emulates the kidney.
- Hypodermic needles have been used since the 1850s, but people's discomfort with needles and the dangers and costs associated with needle disposal have prompted the development of prick-free, painless microfluidic needles ("microneedles").
- Female mosquitoes have developed a complex microneedle organ called a *fascicle* that allows them to extract blood after stinging an animal.
- The tiny volume of saliva used by the mosquito to prevent blood clotting can carry parasites that can cause serious illnesses such as malaria. More than 700,000 people die annually from mosquito bites.

14

The Sensor That Saved Jeff's Feet

Microfluidic Glucose Sensors for Diabetics and Point-of-Care Blood Analyzers

My friend Jeff Spencer, a Hollywood screenwriter and a born storyteller, told me about how he was diagnosed with type 1 diabetes at three years of age in 1969. His doctor issued a cruel warning: "You are going to go blind, and you are going to lose your feet" (p. 97).[1] Jeff did not realize it then, but he was saved by microfluidics and now takes brisk one-hour walks three or four times a week near his home in sunny Los Angeles. Luckily for him and millions of other diabetic patients, there has been a revolution in microfluidic glucose sensors in his lifetime.

Diabetes—technically termed *diabetes mellitus*—can cause life-threatening complications if left untreated, including chronic kidney disease, stroke, coronary heart disease, foot ulcers, and damage to the nervous system and eyes. Diabetes is characterized by the body's inability to adjust glucose levels in the blood. The pancreas is the organ responsible for glucose regulation and balance. When blood glucose is too high, the pancreas secretes more insulin. This hormone tells your cells to absorb glucose from the blood.[a] Glucose is necessary to provide energy to the cells, but high blood glucose levels can damage our bodies. In diabetic patients, either the pancreas does not make enough insulin (the so-called type 1 diabetes) or the body cannot normally respond to the insulin made by the pancreas (type 2 diabetes). Both scenarios cause blood glucose levels to rise. The disease affects approximately 400 million people worldwide, which has been increasing steadily in the past decades due to changes in lifestyle and diet.

The care of diabetics has been recorded throughout history. The Egyptians wrote about the disease in the famous Ebers papyrus dated from 1550 BC. They used the attraction of ants to the high sugar in a diabetic's urine to diagnose it.[2] Indian physicians called it *madhumeha* ("honey urine").[3] In his

[a] The pancreas releases another hormone, glucagon, when blood glucose drops. The hormone tells the liver to release its stored glucose into the blood.

[1] Folch, A. *Hidden in Plain Sight: The History, Science, and Engineering of Microfluidic Technology.* MIT Press (2022).

[2] Rajendran, R., & Rayman, G. Point-of-care blood glucose testing for diabetes care in hospitalized patients: An evidence-based review. *J. Diabetes Sci. Technol.* 8: 1081–1090 (2014).

[3] Lakhtakia, R. The history of diabetes mellitus. *Sultan Qaboos Univ. Med. J.* 13: 368–370 (2013).

Figure 14.1 Portrait of Ibn Sīnā on a 1950 Iranian postage stamp.
Image source: Wikipedia. https://en.wikipedia.org/wiki/ Avicenna#/media/File:1950_%22Avicenna%22_stamp_of_Iran. jpg. Public domain.

five-volume encyclopedia of medicine, *The Canon of Medicine*, the great Persian physician and philosopher Ibn Sīnā (Figure 14.1) described in 1025 AD the sweet taste of diabetic urine and the first treatment that lowered sugar levels.[b,4] In 1675 AD, the English physician Thomas Willis added the Latin term *mellitus* ("honeyed") to "diabetes" after tasting the sweetness of diabetic urine. The transition to a less subjective approach to the analysis of diabetic urine began with the British physiologist Matthew Dobson in 1776.[5] He drew blood from diabetic patients, separated the blood cells from the liquid part of the blood (called *plasma*), and reported that, like the urine, the plasma also had a sugary taste.[c] Less than a century later, in 1850, a Parisian chemist named Edme Jules Maumené created a simple method to detect high sugar in urine using a *redox reaction*.

When two molecules react by exchanging one or more electrons, chemists say that the molecule that has gained electrons has been *reduced*, and the molecule that has lost electrons has been *oxidized*. Because the oxidation and the reduction processes are entangled, the reaction is named a "redox" (short for reduction–oxidation) reaction. We can imagine the process as a handshake: The reducing agent agrees to give electrons to the oxidizing agent and only to that molecule. In some cases, heat is necessary to bring the two molecules to a point where they can agree to make the handshake happen.

Maumené's innovation involved nineteenth-century microfluidic technology. He soaked a strip of sheep wool with tin chloride by capillarity and, once dried,

---

[b] Ibn Sīnā (980–1037 AD) is more commonly known in the West as Avicenna. His *Canon of Medicine* set the standards for medicine in Medieval Europe and the Islamic world and remained a medical authority for centuries. It was used as a standard medical textbook through the eighteenth century in Europe. His treatment for diabetes used a mixture of lupine, fenugreek, and white turmeric—a treatment that lowers sugar levels and is still prescribed in some areas of the world.[4]

[c] Dobson also isolated the brown crystals from their dried urine, identified them as sugars, and found them to taste like brown sugar.[5]

[5] Moodley, N., Ngxamngxa, U., Turzyniecka, M. J., & Pillay, T. S. Historical perspectives in clinical pathology: A history of glucose measurement. *J. Clin. Pathol.* 68: 258–264 (2015).

soaked the strip with urine to detect whether the patient had high sugar in their urine. The wicking action of the fluid (see Chapters 11 and 12) brought the urine (and its sugar) into contact with the tin chloride. The strip would turn black upon heating if it had a high sugar content.[6] In Maumené's reaction, sugar in the urine was the reducing agent being oxidized, and tin chloride was the oxidizing agent being reduced. To be more precise, the tin chloride soaked up by the wool strip reacted with a specific chemical group called an aldehyde group present in sugars such as glucose. Glucose is a molecule formed of a ring of six carbon atoms with an aldehyde group tail. Upon heating, the aldehyde group in the glucose of Maumené's patients' urine became oxidized and reduced the tin chloride to a metallic-black tin.

Maumené's chemical method persisted as the standard test for more than a century until methods based on glucose oxidase—a natural molecule that reacts specifically with glucose and oxygen—were introduced in the 1950s.[5] Maumené's chromogenic (color-generating) reaction had two drawbacks. First, other sugars, such as fructose, also contain an aldehyde group. Thus, their oxidation reaction was not specific enough to signal the presence of glucose alone. Second, it was not quantitative; in other words, the doctor could not easily infer the sugar concentration in the urine from the coloration. In 1957, the Ames Corporation launched a quantitative glucose urine-testing strip infused with glucose oxidase that specifically reacted with glucose[7] and an optical reader.[8] But the measurement from these platforms—as is often the case with optical readouts in non-laboratory settings—was susceptible to human error.[5]

Jeff has followed the developments of glucose monitoring technology with the same meticulous attentiveness that he devotes to his plots and characters—so my conversation with Jeff quickly turned into an invaluable lecture on the history of glucose meters. He first relied on color-changing paper strips for monitoring his blood sugar at home in the 1970s. "As a boy, I remember peeing on a paper stripe, and when it turned green, my mom had me run around the block to burn the extra carbs—that's all we had back then," he recalled (p. 99).[1] The first instruments with direct electrical readouts used large electrodes that produced an electrochemical reaction with the blood. These were not portable and required many operator-dependent steps. Initially, they were neither precise nor user-friendly enough to be left in the patients' homes for them to interpret the results.

These first blood glucometers were in hospitals, providing infrequent blood analyses for the average diabetic. Jeff recalls with terror his biannual glucose

    [6] Maumené, E. J. Sur un nouveau réactif pour distinguer la présence du sucre dans certains liquides. *J. Pharm.* 17: 368–370 (1850).
    [7] Free, A. H., Adams, E. C., Kercher, M. L., Free, H. M., & Cook, M. H. Simple specific test for urine glucose. *Clin. Chem.* 3: 163–167 (1957).
    [8] Kohn, J. A rapid method of estimating blood-glucose ranges. *Lancet* 273: 119–121 (1957).

monitoring visits in the 1970s at the Milwaukee County Children's Hospital near Lake Michigan:

> Twice a year, three times in a row, the day of the scheduled visit, all ten of us diabetic kids were lined up and watched the next kid being punctured in the fingertip with a sharp blade that felt like a can opener on the skin, and they took the blood into the lab. It looked like a camping accident. (p. 101)[1]

Unfortunately, health care providers used to believe that patients could not operate blood glucometers by themselves at home.

Richard Bernstein, an engineer and a diabetic patient who, like Jeff, had type 1 diabetes, proved the health care providers wrong.[9] He was starting to develop complications of diabetes, and whenever he missed a treatment, he would have a hypoglycemic attack (too little sugar in the blood). Fortunately for him, Bernstein's wife was a doctor and managed to get him a machine for home so he could monitor his blood sugar himself and adjust his insulin accordingly. Thanks to his self-monitoring, Bernstein's frequent hypoglycemic attacks were resolved, and his health improved. Bernstein personally convinced Ames Corporation to try to market the machine to patients.

Although the procedure worked for Bernstein, it was not user-friendly. Jeff remembers the first commercially-available version of the machine: "It was the size of a brick. I used it for a month twice a day and then stopped using it" (p. 101).[1] The test was cumbersome, a five-minute process that involved a calibration procedure and a critical, harrowing ten-second delay, among other steps.

Still, the journals refused to publish Bernstein's studies. He was so determined to demonstrate that blood glucose could be self-monitored by patients that, in his forties, he entered medical school, specialized in endocrinology, and started his own diabetic clinic. Using patients from his clinic, he published the first studies on self-monitoring in the 1980s, which started the trend of allowing patients to monitor their sugar levels.[9] In the 1990s, manufacturers finally became interested in offering smaller, more reliable, and more user-friendly glucometers. As a result of the development of microelectronics, they could package the "brick" into a small handheld reader (Figure 14.2). At the same time, they miniaturized the sensing electrodes to use a disposable microfluidic strip that wicked in a drop of blood from a finger prick (the kits included a lancing device) for analysis.

* * *

[9] Bernstein, R. K. Blood glucose control. *Arch. Intern. Med.* 141: 267–268 (1981).

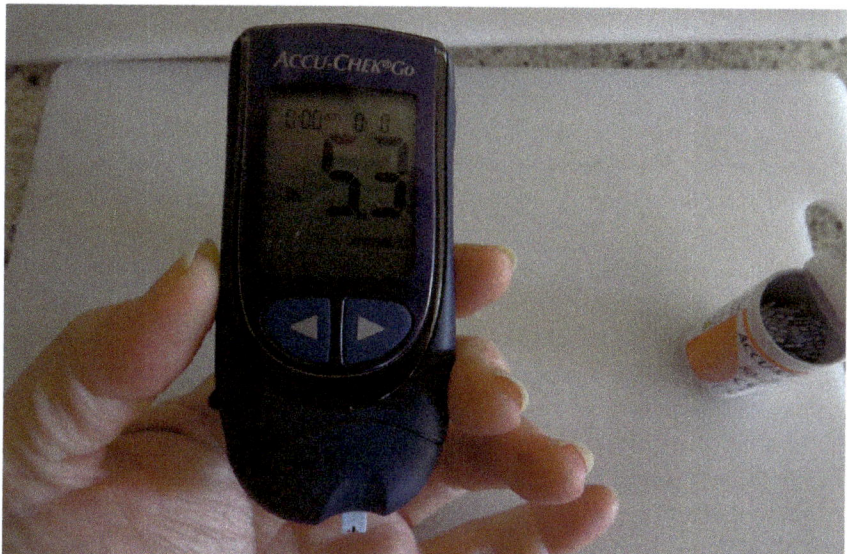

Figure 14.2 Microfluidic glucometer. The user first needs to draw blood with a lancing device or a needle. When the strip contacts the blood drop, blood wicks into the strip by capillarity. The strip (white, bottom) is inserted into the glucometer. Inside the strip, electrodes loaded with an enzyme perform an electrochemical reaction in a few seconds and report the result to the handheld electronic reader. The strip is disposable.

Image source: Pxhere.com. https://pxhere.com/en/photo/867711

Blood analysis by doctors can reveal so much more beyond glucose levels. Blood delivers nutrients and oxygen to cells and transports metabolic waste products away from those same cells. Cells, proteins, and hormones in blood meander through our bodies to regulate our well-being, continuously fighting invaders and fine-tuning our temperature, hunger, thirst, and libido. Hence, engineers have focused much effort on designing apparatuses that analyze multiple parameters from blood, both for clinical and home use.

The development of commercial microfluidic blood analyzers stands as one of the most fascinating pursuits in the history of health care technology. If you have ever been to a hospital and the nurse or the doctor came back with the results of a blood analysis within ten minutes of drawing your blood, the analysis was likely performed with an i-STAT[10] or similar device. The i-STAT was the first of these analyzers, a revolutionary and highly successful portable

---

[d] The i-STAT was also the first health care product to incorporate microfluidic chip technology.

[10] Lauks, I. R. Microfabricated biosensors and microanalytical systems for blood analysis. *Acc. Chem. Res.* 31: 317–324 (1998).

device invented by British chemist-turned-engineer and entrepreneur Imants (pronounced "*Ee-mants*") Lauks in 1984.[d]

Some have touted Lauks as a sort of king Midas of microfluidics, a pioneer who left the security of academia because he was firmly convinced he could better develop his groundbreaking ideas in the industry—and turned them into gold. He has both a carefree resemblance to Richard Gere and an impenetrable aura. To protect his privacy, and perhaps also to reinforce his now legendary reputation, he has conceded very few interviews over the years—so I was pleasantly shocked that he answered my call.

Nothing in his early life would lead you to predict that this man would go this far. Lauks was born to a modest family in 1952 in Bradford, a small Industrial Revolution boomtown in Northern England close to Leeds and Manchester. His parents, young World War II Latvian refugees, had been sent to Bradford, and his father was given a job in the coal mining industry. But the young Lauks lad grew into a bright student who studied chemistry at the Imperial College London and then continued for a PhD in electrical engineering at the same institution. "I was an early exponent of multi-disciplinarism—that was the toolset in my education that gave me good qualifications to do the kind of things that I've been doing," he reasoned. He emigrated to the United States in the 1970s. A rising star, by thirty years of age he had already become a tenured professor of electrical engineering at the University of Pennsylvania. Then, he decided to leave academia to pursue his dream.

Lauks realized that for health tests, the value of a result rapidly diminishes with the time elapsed since the doctor has taken the test. Hence, it would be highly valuable to have an instrument that produces multiple measurements simultaneously, as close to real time as possible, and as near the patient as possible. If he could realize that vision, it would be possible to move away from centralized—sluggish and expensive—hospital testing facilities and enable much faster and cost-effective patient-centered testing by the bedside. The then-revolutionary strategy, called point-of-care testing, has been adopted in almost every major hospital worldwide to complement the centralized testing system. This vision motivated him to leave the university system, move to Canada, and found the startup i-STAT in 1983.[11] He was thirty-one years old. Based in Ottawa, Eastern Canada, i-STAT sought to develop the first commercial microfabricated biosensor and general-purpose handheld blood analyzer.

The i-STAT was groundbreaking at the time. When Lauks joined the University of Pennsylvania, he had become aware of the latest research on

---

[11] Success is in the blood: The toil and sweat of biotech pioneer Imants Lauks. (2010).

chemical sensors by Piet Bergvald in Holland.[12-14] Bergvald had repurposed transistors by coating their electrode tips with polymers so that the current could detect particular ions in contact with the surface of the transistor. Lauks saw an opportunity and improved the fabrication process of Bergvald's ion-sensitive transistor. To insulate the electrodes along the rest of their length, the i-STAT engineers coated them with a light-sensitive polyimide polymer. They next patterned the polyimide with ultraviolet light to expose only the tip of each electrode. The engineers deposited different ion-selective polymer films at the electrode tips using an innovative pneumatic dispenser tool mounted on a precise positioner—much like some modern 3D printers, but long before 3D printers existed.

One of the unique strengths of Bergvald's transistor and the i-STAT was that they borrowed molecules from Nature to detect medically relevant ions in the blood. To endow polymers with ion selectivity, the engineers mixed plastics with substances that act to carry specific ions across the membrane of cells in Nature. An example of one of these substances, called ionophores, is valinomycin, which carries potassium ions across the membrane of bacteria. When mixed with the plastic polymer in a microscopically thin film, valinomycin—although no longer in the bacterial membrane but surrounded by polymer chains—makes the plastic thin film selectively permeable to potassium. In addition to potassium, they created films that were selective for many other ions: sodium, chloride, calcium, hydrogen, and ammonia ions. An array of these electrodes on the silicon chip could detect various ions inside a small chamber.

Bringing fluids reliably to the sensor was another challenge. "To me, the real innovation was the original design of the fluidic architecture," Lauks said. The chamber and microchannel were fabricated in plastic by injection molding,[10] a technique that involves melting and flowing plastic into a metal mold at high pressures. The channel had a cross-section of approximately 1 by 1 millimeter and was 30 millimeters long. Compared to the glucometer strips, which only measured glucose concentration, the i-STAT cartridge was much more demanding and complex to manufacture, but it could measure many blood substances at the same time. The initial system integrated an electrochemical chip (inside a cartridge) and a handheld reader with a pneumatic actuation system to displace liquids across the chip. This system performed six tests (calcium, sodium, potassium, dissolved carbon dioxide and oxygen concentrations, and the number of red cells per unit volume or hematocrit) in a small channel in ninety seconds—instead of fifteen to twenty minutes for larger analyzers.

[12] Bergveld, P. Development of an ion-sensitive solid-state device for neurophysiological measurements. *IEEE Trans. Biomed. Eng.* 17: 70–71 (1970).
[13] van der Schoot, B. H., & Bergveld, P. An ISFET-based microliter titrator integration of a chemical sensor–actuator system. *Sensors & Actuators* 8: 11–22 (1985).
[14] van der Schoot, B. H., & Bergveld, P. The pH-static enzyme sensor: An ISFET-based enzyme sensor, insensitive to the buffer capacity of the sample. *Anal. Chim. Acta* 199: 157–160 (1987).

The i-STAT now comprises twenty-five different tests performed in just ten minutes. The user only has to insert a cartridge into a handheld reader, and the electronic screen displays the measurements in easy-to-read numbers. Lauks left i-STAT in 1999 to start a competing, cheaper product, the *epoc* (Figure 14.3), which simultaneously measures the concentration of various essential ions (sodium, potassium, and calcium), the acidity of the blood, the concentration of oxygen and carbon dioxide, and, of course, glucose.

* * *

The most modern glucometers are wearable, working to make the lancing kits a thing of the past. My friend Jeff presently wears a microfluidic needle skin patch that pierces through the skin of his arm, providing continuous sampling of his

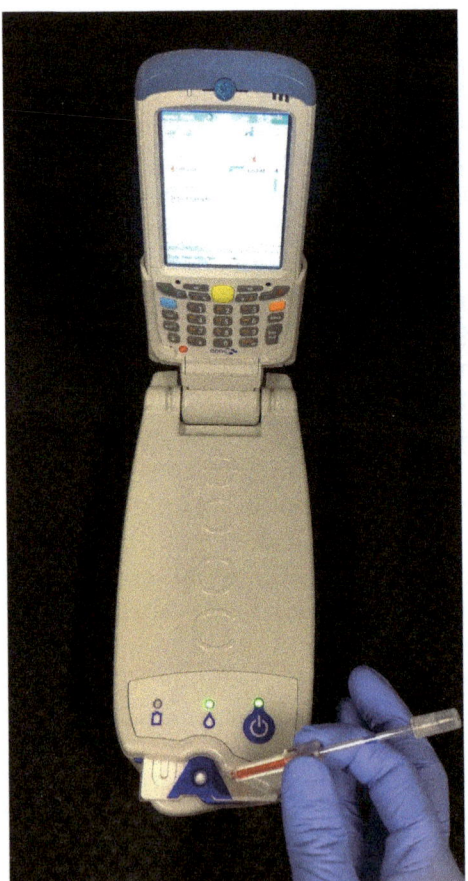

Figure 14.3 The epoc, a wireless and more affordable version of the i-STAT. The microfluidic cartridge is shown inserted at the bottom, as it is being loaded with test solution by the user.

Image courtesy of Mark Wener, Department of Laboratory Medicine & Pathology, University of Washington.

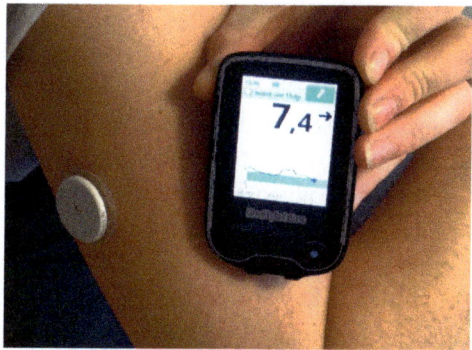

Figure 14.4 A continuous glucose monitor. The microneedle sensor (white circle) is fixed to the upper arm and scanned with the reader. The reader is showing (top to bottom) time (13:46), days to replacement of sensor (11), current blood glucose in millimoles per liter (7.4), and a diagram of the latest blood glucose levels.

Image source: User Sjö on Wikipedia. https://en.wikipedia.org/wiki/Glucose_meter#/media/ File:BGM_twopart.JPG (BY-SA 4.0).

blood glucose levels (Figure 14.4).[15] The microneedles are so tiny that they are painless—says the commercial; Jeff disagrees and says, "You feel it, but I can't complain"—as he does not need to prick his finger anymore. Using Bluetooth, the patch "speaks" to a smart insulin pump he wears. Together, they coordinate the control of his blood sugar with insulin. In addition, the device helps prepare his body for sugar spikes from food. After dialing "65 units" of carbs, he can safely eat a bagel, so his insulin pump anticipates his blood glucose rise. Thanks to an automatic regulation mode, the line on the screen of his handheld monitor displaying his glucose levels, updated every few minutes, stays relatively flat—as it would be in a non-diabetic person—and he no longer sees wild fluctuations.

These microfluidic devices and improvements are a dream come true for blood analysis, delivering fast and accurate measurements that impact millions of people worldwide. And they disproved the heartless predictions of Jeff's childhood doctor: Jeff did not go blind, and he still had both feet the last time we took Muppet, his dog, for a walk.

Summary

- Diabetes is a disease characterized by the body's inability to adjust the levels of glucose in the blood.

---

[15] Heller, A. Implanted electrochemical glucose sensors for the management of diabetes. *Annu. Rev. Biomed. Eng.* 1: 153–175 (1999).

- The first urine strips for detecting diabetes were invented by Edme Jules Maumené in 1850. Capillary wicking of urine started a chemical reaction (called a *redox reaction*) between a chemical on the wool strip (tin chloride) and sugars present in the urine of the patient.
- In the 1950s, the Ames Corporation introduced urine and blood strips that were more specific to glucose because the chemical reaction for the detection of glucose was based on glucose oxidase, a natural enzyme. Many patients used these simple microfluidic devices to monitor their glucose levels; however, they were susceptible to human error.
- In 1984, British entrepreneur Imants Lauks invented the i-STAT, a miniature blood analyzer based on electrochemical sensing using a drop of blood in a microchannel.
- All modern glucose meters (*glucometers*) are essentially simplified i-STATs, microfluidic devices that incorporate the enzyme glucose oxidase to start the electrochemical reaction.
- The latest continuous glucose monitors sample the patient's blood continuously with microfluidic needles ("microneedles") that pierce the skin and transmit the data to a wearable glucometer unit, thus bypassing the need for lancing the skin.

15

Every Breath We Take

The Microfluidic Organs That Breathe for Us and the First Lung-on-a-Chip

Craig Foster, a South African nature filmmaker, had just completed a groundbreaking film about the San bushmen of the Kalahari Desert and felt utterly burnt out. To regain his connection to Nature and his own family, he pledged to free dive—no wetsuit and no oxygen—365 days a year for ten years in the icy waters off his backyard in the great African kelp forest of the Cape Town Peninsula.[1]

Foster's frequent observations enabled him to see and experience things others had missed. He said, "This crazy idea was in my mind for a long time and then eventually I started seeing the first underwater tracks, that's when I first thought it could work but I had no idea that I could develop it into such a detailed way understanding of animals underwater."[2] One day, he spotted a strange ball of shells and rocks which he realized was an octopus protecting herself with a quickly assembled camouflage (Figure 15.1)—a behavior he calls "armoring." In his daily dives, he saw her hunt and play with fish, and he eventually *befriended* her. This remarkable friendship, which led to an Oscar-winning documentary about his experience called *My Octopus Teacher* (Figure 15.2),[2] may not have been possible without his ability to swim underwater for extended durations—six minutes—on one breath of air. And behind every breath he and every other animal take—octopus, whale, or zebra—there lies an intricate network of air-filled microfluidic channels.

What are the limits of human breath? When you plunge into the blue silence without assistance other than that last gulp of air, everything seems to stop against the oxygen-ticking clock: noise, gravity, even time stands still for a bit. Until the burning sensation in the lungs calls you back to the surface. Humans have been diving for thousands of years, attracted by the challenge of entering a viscous, weightless environment and meeting the ocean's exotic

---

[1] Stark, V. "Octopus teacher'" lets filmmaker into secret world. VoA News. https://www.voanews.com/a/arts-culture_octopus-teacher-lets-filmmaker-secret-world/6195670.html (2020, September 9).

[2] Allie, M. "How I became friends with an octopus." BBC News. https://www.bbc.com/news/world-africa-45967535 (2018, November 3).

Figure 15.1 Scene from *Planet Earth: Blue Planet II*, a BBC America documentary. The ball of shells at the center of the image is an octopus in disguise.
Image source: YouTube/BBC. https://www.youtube.com/watch?v=GoTk5WofgoE. Fair use.

Figure 15.2 Scene from *My Octopus Teacher*, Craig Foster's documentary.
Image source: Netflix. Fair use.

biological variety. So, it's no surprise that diving records are broken very often. Record holders such as Mateusz Malina, who swam the record underwater distance of five Olympic-size pools—250 meters—in 2022, maximize time below the surface by minimizing exertion. Malina does not swim frantically; instead, he glides calmly to keep his heart rate down. As of 2023, Branko Petrović holds the world record time for a non-oxygen-assisted breath hold at 11 minutes and

54 seconds; for women, Natalia Molchanova holds the record at 9 minutes and 2 seconds. Others breathe in pure oxygen, lie completely still in freezing water, and stay underwater even longer to see how far the human body can go. Still, there is always an inescapable physiological limit.[a]

Maybe more impressive are those who hold their breath as part of their occupation. In normal conditions, when oxygen consumption is higher, humans can stay underwater for only a few minutes. In intense—some would say bizarre— sports such as underwater rugby, underwater football, or underwater hockey, the average duration of the dives by the players—who wear a snorkeling mask and fins but otherwise use only their lungs to breathe—is less than a minute. The *Ama* (Figure 15.3) are Japanese female free divers who carry on a centuries-long tradition of catching coveted pearls, seaweed, abalone, and other shellfish. They can dive for up to two minutes at a time, 100–150 times a day, due to a lengthy training by their parents that starts in their teens.[3]

Figure 15.3 An Ama diver.

Image source: Yoshiaki Sugaya/Shutterstock. https://www.shutterstock.com/image-photo/ama-women-divers-ancient-japanese-tradition-2300349877.

---

[a] Normal atmosphere contains 21% oxygen, so breathing (and hyperventilating on) 100% oxygen for up to half an hour before completing the breath hold raises the concentration of oxygen in the blood. Using this method, Budimir Šobat obtained the men's underwater world record at 24 minutes and 37 seconds and Karoline Mariechen the women's record at 18 minutes and 32 seconds. Famously, and without prior training, the actress Kate Winslet was able to hold her breath for 7 minutes and 15 seconds during filming of the movie *Avatar: The Way of Water* using 100% oxygen.[4]

[3] Lim, E. The breath-hold of Japan's pearl diving mermaids. *Divers' Digest.* https://www.uw360.asia/the-breath-hold-of-the-pearl-diving-mermaids-of-japan (2021, August 27).

[4] Bain, A. The science of holding your breath: How could Kate Winslet stay underwater for over 7 minutes in *Avatar 2*? The Conversation. https://theconversation.com/the-science-of-holding-your-breath-how-could-kate-winslet-stay-underwater-for-over-7-minutes-in-avatar-2-198381   (2023, February 5).

We have learned a lot about breathing by looking at the animal kingdom. The ultimate "winners" of the breath-holding contest are, not surprisingly, air-breathing marine animals that have adapted to living in cold water. Many aquatic animals with lungs can hold their breath *inside the water* much, much longer than humans.[b] The loggerhead turtle holds the record of living under-water on one gulp of air for up to ten hours.[5] The Cuvier's beaked whale, the deepest-diving mammal, can swim underwater for almost four hours without resurfacing. Among reptiles, marine iguanas hold the record of stay-ing underwater for thirty minutes. The emperor penguin can be under the water surface for more than eighteen minutes, longer than any other bird species.[7]

Octopuses and some fish species would beat any human at "holding their breath," although they do it *out of the water* and might be disqualified in a competition because they get help from breathing extra air through their skin. Skin-breathing is not as efficient as gills, but it allows some octopuses to sur-vive twenty to thirty minutes outside the water, despite not having lungs. The breathing abilities of octopuses have intrigued humans for a long time. Octo-puses are spineless, very primitive animals that roamed the seas long before the dinosaurs—probably around 330 million years ago[8]—but their nervous sys-tems are composed of 500 million neurons, comparable to that of a dog. They can navigate mazes, unscrew caps, dismantle machinery, quickly grab shells to improvise a camouflage that confuses predators,[9] and escape from aquariums after crawling across the floor.[10] People often attribute these abilities to their rel-atively large "brains," but it is often their—less appreciated—ability to breathe outside the water that enables them to pull off human-like pranks. In 1875, aquarists in Brighton, UK, found that the octopuses were sneaking out of their

[b] Humans have a very energy-hungry brain compared to most animals. Despite representing only 2% of the body weight, the brain consumes approximately 20–25% of the oxygen taken in by the lungs.[6]

[5] Shields, J. What animal can hold its breath longest? *Howstuffworks.* https://animals. howstuffworks.com/animal-facts/what-animal-can-hold-its-breath-longest.htm (2024).

[6] Raichle, M. E., & Gusnard, D. A. Appraising the brain's energy budget. *Proc. Natl. Acad. Sci. USA* 99: 10237–10239 (2002).

[7] AnimalFunFacts.net. Animals that are best at holding breath underwater. https://www. animalfunfacts.net/animal-records/195-animals-that-are-best-at-holding-breath-under-water. html (n.d.).

[8] Whalen, C. D., & Landman, N. H. Fossil coleoid cephalopod from the Mississippian Bear Gulch Lagerstätte sheds light on early vampyropod evolution. *Nat. Commun.* 13: 1107 (2022).

[9] Foster, C. *My Octopus Teacher.* https://www.documentaryarea.com/video/My+Octopus+ Teacher (2020).

[10] Brulliard, K. Octopus slips out of aquarium tank, crawls across floor, escapes down pipe to ocean. *The Washington Post.* https://www.washingtonpost.com/news/animalia/wp/2016/04/ 13/octopus-slips-out-of-aquarium-tank-crawls-across-floor-escapes-down-pipe-to-ocean (2018, April 13).

Figure 15.4 Scene from *Extraordinary Octopus Takes to Land*, BBC documentary. *Image source: YouTube/BBC. https://www.youtube.com/watch?v=ebeNeQFUMa0. Fair use.*

aquariums at night to feast on the fish bowls and furtively returning to their dens after their briny bacchanal.[11] This behavior is not limited to octopuses in captivity. At least one species in the Philippines, Indonesia, and southern Australia routinely crawls on land between tidal pools to forage for crabs (Figure 15.4). And virtually all octopuses can move to some extent between two nearby bodies of water.

Evolution has come up with three solutions to the problem of distributing oxygen to all the cells of multicellular animals—and all three of them are microfluidic. The type of solution depends on the size of the organism and the medium in which it lives. Small organisms such as plankton and insects can rely on *diffusion* to obtain oxygen because of the tiny distances involved (Figure 15.5). The oxygen we breathe is a miniscule molecule made of only two oxygen atoms (O_2), so it can bounce through the soup of molecules all tissues are made of until it drifts to its destination—as long as the distance it has to travel is less than a few millimeters. The problem that multicellular animals encountered during evolution was that diffusion was very inefficient for considerable distances. It takes molecules four times as long to diffuse twice as far (recall Fick's law in Chapter 1 and Figure 1.4): a molecule of oxygen needs an average of 8.5 minutes to diffuse 1 mm in water at 20°C, four times as long to diffuse 2 mm, sixteen times longer to diffuse 4 mm, and so on. For larger distances,

[11] Plante, C. Octopus escapes laboratory, continuing long tradition of octopi outsmarting humans. *The Verge*. https://www.theverge.com/2016/4/13/11422148/inky-octopus-escape-national-aquarium-new-zealand (2016, April 13).

Figure 15.5 Planktons are so tiny that they do not need gills or lungs to breathe. Most of the ones in the picture are of the larger type called zooplankton and are approximately 1 millimeter in length. Diffusion of oxygen through their skin is enough for them to stay alive.

Image source: User Matt Wilson/Jay Clark (NOAA NMFS AFSC) on Wikipedia. https://en. wikipedia.org/wiki/Zooplankton#/media/File:Mixed_zooplankton_sample.jpg. Public domain.

the bounce-and-drift trip comes to a halt. The times are even longer when oxygen needs to diffuse through tissue (made mainly of proteins and cells), so you can see that breathing by diffusion is not a very efficient method unless the organism is tiny. Insects do not have neither lungs nor blood (only a fluid called *haemolymph* that sloshes around in their body), but larger insects such as moths and flies are perforated inside with tubes running through their bodies that allow oxygen to diffuse to the inner parts.

Like an experienced microfluidic engineer, Nature seems to be able to ingeniously reach out into its microfluidic tool chest whenever it needs to evolve a new solution.[c] For vertebrates like us, other mammals, birds, amphibians, reptiles, and fish, evolution "came up" with blood to transport molecules over long distances. Blood—essentially a carrier of nutrients, oxygen, and carbon

---

[c] It only *appears* to us to be designed with microfluidic intelligence—Nature only "designs" by trial and error, and over millions of years. The fact that this vastly prolonged trial-and-error process has converged into so many microfluidic solutions only speaks of the high efficiency of microscale fluidic phenomena compared to the macroscale. Indeed, Diffusion works optimally as a way to move substances dissolved in liquids over tiny distances but poorly over large distances, and surface tension is a very strong force when it comes to powering the motion of small fluid volumes but an irrelevant force when dealing with large fluid volumes.

dioxide—is actively pumped by one or more *hearts* and is distributed through a massively parallel network of microfluidic channels—called *blood capillaries*— to load and unload the red blood cells with oxygen (and, in parallel, carbon dioxide).[d] This gas exchange works differently for underwater and terrestrial animals. Large, underwater organisms have *gills* that extract oxygen from water and transfer it into the blood. Large terrestrial organisms—such as all mammals, reptiles, and birds—have *lungs* that directly transfer oxygen from air into the blood.

The evolutionary origins of gills and lungs are connected. Gills appeared approximately 500 million years ago to help aquatic animals transition from small, "worm-like" animals to larger, "fish-like" animals. Interestingly, at the beginning these proto-gills served a different purpose—to regulate the exchange of ions, also originally fulfilled by the skin. It was only later, as vertebrates evolved to become larger, that gills became the body site[12] for gas exchange as well.[13] About 400–375 million years ago, some fish started developing limbs or reptile behaviors that allowed them to explore land.[14] Shortly after this water-to-land transition, some of these creatures evolved an adaptation of the gills to air—the lungs.[15,16]

But Nature is like a bricolage (microfluidic) engineer who likes to try everything at its disposal, so not all animals strictly conform to just one of these three types of respiration (diffusion-, gill-, or lung-based). A particular group of organisms called *amphibians*, such as the frog and the salamander, are born in the water with gills. In their adult life, the gills degenerate, and they start breathing by diffusion *through their skin*, like insects do. Most adult amphibians also develop lungs, but a few, like the lungless salamander, only breathe through their skin. This skin respiration is enough for a frog to stay at rest underwater for several hours, but it needs to surface for air if it expends extra energy swimming. Watching Mateusz Malina leisurely frog-stroking toward his underwater swimming world record makes one wonder in amusement whether he, too, can breathe through his skin. Indeed, skin breathing is not an exclusive attribute of insects and amphibians. Perhaps the most striking example in the whole animal

---

[d] Nearly 1.5% of the oxygen and around 7–10% of the carbon dioxide are directly dissolved in the bloodstream's plasma, even in the absence of the red blood cells.[12]

[12] LumenLearning. Gas exchange. SUNY–OER Services. https://courses.lumenlearning.com/suny-ap2/chapter/gas-exchange (n.d.).

[13] Sackville, M. A., Cameron, C. B., Gillis, J. A., & Brauner, C. J. Ion regulation at gills precedes gas exchange and the origin of vertebrates. *Nature* 610: 699–703 (2022).

[14] Pennisi, E. Genes for life on land evolved earlier in fish: Hidden genetic pathways behind land–water transition are found in living fish. *Science* 371: 658–659 (2021).

[15] Hoffman, M., Taylor, B. E., & Harris, M. B. Evolution of lung breathing from a lungless primitive vertebrate. *Respir. Physiol. Neurobiol.* 224: 11–16 (2016).

[16] Cupello, C., Hirasawa, T., Tatsumi, N., Yabumoto, Y., Gueriau, P., Isogai, S., et al. Lung evolution in vertebrates and the water-to-land transition. *Elife* 11: e77156 (2022).

kingdom is found in the shallow, muddy waters of Australia, Africa, and South America: *lungfish*, a 300-million-year-old species of fish with gills and lungs that can also breathe through their skin.[17] Lungfish need to gulp air like us, but unlike us, they can spend four or five days between gulps with the help of their gills and skin breathing. And some eel species, which generally live their daily lives breathing through their gills and slithering underwater, can undergo extensive overland migrations lasting hours because they can assimilate the oxygen that diffuses through their skin when outside the water[18]—just like octopuses do. Over the ages, Nature has shown it is capable of the weirdest and most beautiful adaptations.

The gill contains sheets of feathery microfluidic channels that allow aquatic animals to extract the small amount of oxygen dissolved in water. Diffusion is horribly slow across long distances, so evolution could not have devised the gills as a big balloon-looking organ. Instead, Nature broke down the gill's cavity into thousands of microfluidic filaments—called *lamellae*—filled with blood inside and exposed to water on the outside (Figure 15.6). This way, gases can be exchanged by diffusion over small distances (like it occurs in plankton) in many places in parallel. The wall of the lamellae is a thin, semipermeable membrane formed of a single layer of cells that acts as a filter.

As oxygen-rich water rushes into the gill, oxygen rapidly diffuses to the blood side of the membrane until the oxygen concentration is equilibrated, oxygenating the blood side and depleting the water from 75% of its oxygen. Carbon dioxide produced by the underwater animal's cells, which ends up in the animal's blood, undergoes the opposite fate: The carbon dioxide diffuses from the blood into the water through the membrane. The high efficiency of the gas exchange, on which fishes' lives depend, would not be possible were it not for the lamellae's small width—less than that of a human hair.[e]

Lungs are to terrestrial animals what the gill is to aquatic animals—equally marvelous, complex, and microfluidic in their every detail. Each time you inhale—right now—air fills the lung, this wondrous 6-liter spongy tree whose hollow microfluidic branches are called *airways*. Each airway ends in a grape-cluster-looking group of microscopic and elastic air sacs called the *alveoli* (Figure 15.7). Tiny blood vessels envelop the alveoli. When you inhale, your 500

---

[e] A number of fishes that live in oxygen-poor water need to come to the surface to gulp air. This extra air is reabsorbed through the lining of internal structures in the pharynx or the stomach by diffusion, so it is a form of respiration by diffusion.[17] Some species, such as the semiaquatic alpine newt, can sometimes have external gills, and others, such as the hermit crab, can breathe outside the water provided the gills are kept moist.

[17] White, F. N., & Burggren, W. W. Dynamics of vertebrate respiratory mechanisms. *Encyclopedia Britannica*. https://www.britannica.com/science/respiratory-system/Dynamics-of-vertebrate-respiratory-mechanisms (2024).

[18] Kean, Z. Eels can travel over land, climb walls and take down serious prey. They may be Australia's most hardcore animal. ABC Science. https://www.abc.net.au/news/science/2022-01-09/eels-australia-most-hardcore-animal/100572614 (2022, January 8).

Figure 15.6 Gill of a tuna fish, showing the lamellae (red).

Image source: ArtDary/Shutterstock. https://www.shutterstock.com/image-photo/gills-tuna-bonita-fishing-pacific-ocean-1670340799.

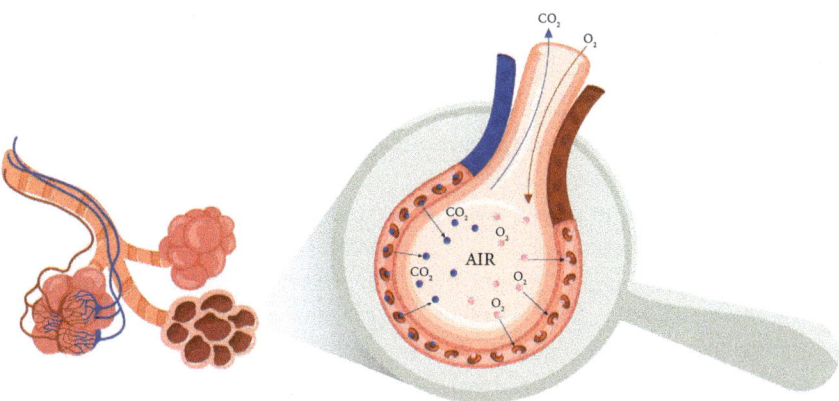

Figure 15.7 The alveolus exchange of oxygen (into the blood capillary) and carbon dioxide (out of the capillary) is a diffusion process that is very efficient because it happens on a microfluidic scale.

Image source: Tatsiana Matusevich/Shutterstock. https://www.shutterstock.com/image-vector/pulmonary-alveoli-oxygen-carbon-dioxide-exchange-2234215625.

million alveoli expand like microscopic balloons, each operating like a microfluidic gas station, where red blood cells reload with oxygen (the "fuel"). The "fuel tank" for oxygen in red blood cells is a molecule called *hemoglobin*, of which each red blood cell contains close to 270 million copies. This molecule acts as a sort of double-sticky tape for both oxygen and carbon dioxide, always preferring the former in case of conflict between the two. The red blood cells are sent out to travel through all the body's capillaries, taking random paths and following the flow, orderly squeezing in single-cell file as they do so. As they pass close to cells that are low in oxygen (and high in carbon dioxide)—for example, your muscle cells when you are running or your brain cells trying to understand this—hemoglobin releases oxygen to power the cells. As soon as the surrounding cells gulp up the oxygen, the hemoglobin becomes sticky to the lingering carbon dioxide and the red blood cells return to the lungs (with the carbon dioxide waste) for another round. When the red blood cells make it back to the alveoli, the blood capillaries are highly oxygenated by the proximity of the alveoli, and hemoglobin prefers oxygen, so it offloads the carbon dioxide in the gas you exhale. The cycle restarts. That's how each of the 30 trillion cells in the human body gets its oxygen and gets rid of its carbon dioxide. It's as if Amazon delivered oxygen bottles to every house on the planet and also collected everybody's trash.

Despite the varied approaches developed by Nature, they all share the same basic underlying mechanism of diffusion. Small organisms such as insects do not need lungs—for their size, diffusion alone does the trick. Larger organisms rely on lungs and gills to breathe, but if you look closely at a cellular scale, all these gas exchanges by red blood cells are based on diffusion. The lung, as with the gills and their lamellae, can breathe for the whole organism because it uses microfluidics to massively parallelize the diffusive gas exchange between millions of tiny alveoli and the blood capillaries. Any way we look at it, oxygen intake in all large multicellular organisms is a highly parallelized microfluidic process—every breath we take.

To better understand how massive that parallelization is, all the airways in an adult human add up to about 2,500 kilometers of tiny conduits.[19] If we could unfold and stitch the inside lining of all the airways together, they would cover an area the size of a tennis court. We fill the lungs 17,000 times a day, a total of more than 7,500 liters of air.[19] That is equivalent to the volume of a small swimming pool and the same volume of blood that the heart pumps in one day. Evolution used microfluidics to optimize the metabolic requirements of our species by precisely designing the ratio of the lung capacity to the blood volume.

---

[19] American Lung Association. How your lungs get the job done. *Each Breath* [Blog]. https://www.lung.org/blog/how-your-lungs-work (2017, July 19).

But lungs do not always function as admirably well as Craig Foster's or the Ama divers' did. Sometimes, we catch a cold, flu, or pneumonia and sneeze, cough, or make wheezing sounds when we breathe. Then doctors such as Jim Grotberg use a stethoscope to *auscultate* (listen to) our lungs. The stethoscope is a simple instrument that is used millions of times a day throughout the world.[f] For more than two centuries, doctors have used stethoscopes[g] to listen to the lungs for wheezing sounds, which are prevalent in emphysema—a condition in which the air sacs are abnormally enlarged due to airborne irritants, leading to breathlessness—and asthma, and the crackling sounds that occur in pneumonia or fluid overloading. These sounds inspired Dr. Grotberg and his colleague, Dr. Shuichi Takayama, to build a microfluidic chip living avatar of the lungs' airways and alveoli to better understand the lungs' function.

* * *

Jim Grotberg seemed pre-destined to become one of the world's authorities on the biomechanics of breathing. He was born in Oak Park, a suburb of Chicago, home to illustrious citizens such as novelist Ernest Hemingway and influential architect Frank Lloyd Wright. Grotberg has followed in their footsteps: The American Society of Mechanical Engineers has cited Grotberg as "the leading biofluid dynamics expert in the United States and among the top two or three in the world."[20]

Grotberg grew up using his lungs a lot. "I played trombone in high school and was selected for the Maryland All-State Orchestra for my sophomore, junior, and senior years," he told me. At 6 feet tall and with a strong build befitting his name—*Grotberg* in Norwegian means "Great Mountain"—he played basketball during high school and with Cornell Big Red, Cornell University's men's basketball team. Inevitably, Grotberg would become interested in studying how the lungs work.

During his internship in anesthesiology at Northwestern University in Chicago in the winter of 1981, Grotberg came across a patient in the intensive care unit who had pneumocystis pneumonia, a severe lung infection caused by a fungus. Grotberg, who specialized in emergency medicine, remembers him very well to this day: "He was thin, with dark hair. He was a member of the Chicago LGBT community, and his sister visited him every day. He also had Kaposi's sarcoma which is a rare skin cancer," he recalls. The patient is now remembered as the first HIV AIDS patient at Northwestern University. Grotberg, who had to

[f] There are approximately 10 million physicians and 20 million nurses in the world (source: World Health Organization).

[g] The stethoscope was invented in 1816 by René Laënnec (1781–1826), a French physician.

[20] Findatopdoc.com. https://www.findatopdoc.com/doctor/995928-James-Grotberg-Emergency -Physician (n.d.).

rotate the patient in agony to his side to listen to his lungs with a stethoscope, retained an essential detail: "He had very crackly lungs."

The fascination he experienced while listening to patients' chests with a stethoscope stayed with him long after his initial training. Grotberg eventually became a doctor–researcher and professor at the Biomedical Engineering Department at the University of Michigan, and he kept wondering about the sounds heard with the stethoscope. What if, Grotberg thought, these sounds were not so much a *sign* of the disease but its *cause*?[21]

Grotberg's main challenge was he could not insert a microscope into the patient's lungs to *see* what was going on. It has been appreciated for centuries now that the lung is the main battleground for respiratory pathogens trying to enter our body. Yet little is known about the effect of critical physiological parameters and how infection occurs at the cellular and tissue levels. For example, could the speed of inhalation or the thickness of the mucosa layer influence how the pathogen attacks? A thin fluid film covers the airways in a healthy lung to moisten the tissue. Under certain conditions—for example, due to inflammation during the attack of a specific pathogen such as a virus—the tissue reacts by producing much more fluid, which causes the airways to narrow. When that happens, there is a higher chance of the airway occluding by a liquid plug moving down the airways until the plug ruptures. Doctors like Grotberg can hear—but not *see*—the plugs rupturing as a crackling sound with a stethoscope and infer how your lungs are doing by listening to them. In particular, Grotberg was trying to prove the hypothesis that the airway-lining cells were *harmed* by the liquid plugs causing the crackling sounds. However, the theory was difficult to prove without building a system that could adequately observe the lung's airways, with cells and all.

Whenever human experimentation has fallen short, most scientists in the past century have used the same approach: use animals in our stead. Researchers have split open the bellies, chests, and skulls of frogs, mice, rats, rabbits, dogs, sheep, pigs, fish, birds, and many other species to learn about human biology and save human lives. Alarmingly, the number of vertebrate animals sacrificed to biomedical research has grown to slaughterhouse proportions[h]—more than 100 million creatures annually, according to some counts.[22] Defenders of what is coming to be viewed as an inhumane practice argue that 95% of the creatures are expendable rodents and that two-thirds of Nobel Laureates have relied

---

[h] Nearly 300 million cows are killed every year for food.

[21] Grotberg, J. B. Crackles and wheezes: Agents of injury? *Ann. Am. Thorac. Soc.* 16: 967–969 (2019).

[22] Taylor, K., Gordon, N., Langley, G., & Higgins, W. Estimates for worldwide laboratory animal use in 2005. *Altern. Lab. Anim.* 36: 327–342 (2008).

on animal data for their research.[23] Animal activists counter that it is cruel and unnecessary because the suffering imparted on these millions of beings does not outweigh the benefit obtained in terms of human health: Animal experiments are very poor predictors of outcomes in human disease.[i,24]

Grotberg had a better idea that involved neither animals nor human subjects but, rather, something in between. Luckily for him, in 1999, a bright chemist-turned-microfluidic-engineer named Shuichi Takayama was finishing his postdoctoral research at Harvard University and interviewed for a faculty position in Grotberg's department. At Harvard, Takayama had kept cells alive inside an oxygen-permeable, transparent-rubber microfluidic channel and used laminar flow to separate reagents by invisible fluid walls above the cells.[25] In his seminar, Takayama lectured about microfluidics and vascular endothelial cells. "It occurred to me that I could get him involved in airway cells and plug flows using cell-lined channels," Grotberg recalls. "So we hired him, and I shared my student Dan Huh to develop the interaction." Conveniently for the newly formed team, Takayama was given an office next door to Grotberg.

However, the traditional microfluidic channel that Takayama brought from Harvard was insufficient to address the sophisticated questions that Grotberg had in mind. The lung's airways are tiny air-filled tubes with walls of just one layer of airway epithelial cells. These cells are extraordinary because they are in contact with moist air on one side and touch blood on the other, constantly receiving nutrients. The cells that Takayama had been culturing in his microchannels at Harvard were sitting on a solid surface and covered with fluid, so they were not exposed to air. Under these unnatural conditions, the delicate human lung epithelial cells would lose some of their essential properties. In biologist's jargon, they would lose their "differentiated" state.

Huh and Takayama came up with a very innovative "sandwich" design that made it possible to re-create the biology of the airway. To imitate the airway

[23] Burggren, W. W., & Warburton, S. Amphibians as animal models for laboratory research in physiology. *ILAR J.* 48: 260–269 (2007).

[i] To help expedite this vastly inefficient process, in December 2022, the U.S. Congress passed the FDA Modernization Act 2.0 that lifts the requirement of animal testing in preclinical studies.[26] The European Parliament has also recently passed a resolution to phase out animal research.[27]

[24] Akhtar, A. The flaws and human harms of animal experimentation. *Cambridge Q. Healthc. Ethics* 24: 407–419 (2015).

[25] Takayama, S., McDonald, J. C., Ostuni, E., Liang, M. N., Kenis, P. J. A., Ismagilov, R. F., & Whitesides, G. M. Patterning cells and their environments using multiple laminar fluid flows in capillary networks. *Proc. Natl. Acad. Sci. USA* 96: 5545–5548 (1999).

[26] Hernandez, J. The FDA no longer requires all drugs to be tested on animals before human trials. NPR. https://www.npr.org/2023/01/12/1148529799/fda-animal-testing-pharmaceuticals-drug-development (2023, January 12).

[27] European Parliament. The use of animals for scientific research in Europe. https://www.europarl.europa.eu/stoa/en/events/details/the-use-of-animals-for-scientific-resear/20220530 WKS04241 (2022, June 28).

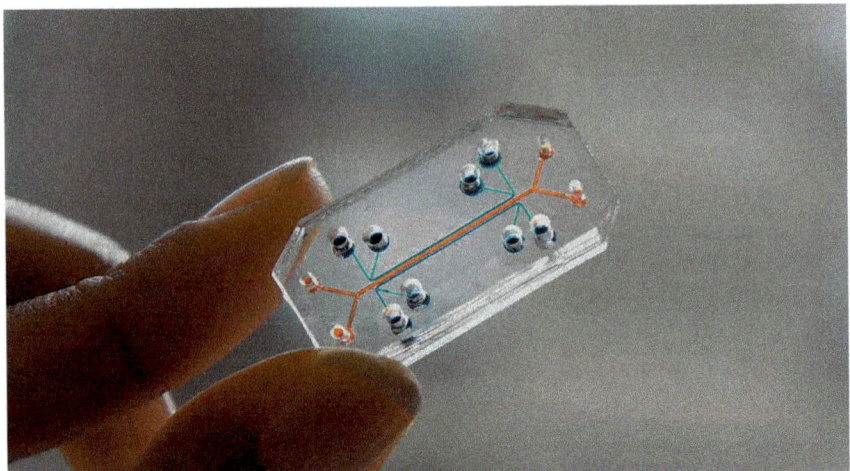

Figure 15.8 This lung-on-a-chip device is composed of a thin porous membrane sandwiched between two transparent rubber layers. The bottom layer forms a fluid microchannel (acting as the "blood") and the top layer forms an air microchannel (acting as the "airway"), while the cell-covered membrane acts as the wall of the alveolus.

Courtesy of Wyss Institute at Harvard University.

function, Huh put a polydimethylsiloxane (PDMS) channel on top, another at the bottom, and sandwiched a polyester porous membrane in the middle (Figure 15.8). These artificial airways had a square cross-section (instead of a round one like the lung ones), but the cells could be seeded on a porous surface (like in the lung) that separated air from the liquid. "That design was key to differentiating cells in a device because we had to culture them under air while feeding them [with fluid] from the bottom," says Huh. "It worked beautifully," he remembers.[28]

Huh seeded commercially available cells in this sandwich design. Next, he used the device to dynamically switch "plugs" of liquid on and off and send those plugs over the cells. They found that the plugs did emulate the crackling. Furthermore, cells would die when the plugs ruptured, "so we were able to claim that crackling might be associated with mechanical lung injury,"[28] Huh adds, describing the conclusion of their 2007 study.[29] This device would be named

[28] Folch, A. *Hidden in Plain Sight: The History, Science, and Engineering of Microfluidic Technology.* MIT Press, 2022, 200, 201, and 203.

[29] Huh, D., Fujioka, H., Tung, Y.-C., Futai, N., Paine, R., Grotberg, J. B., & Takayama, S. Acoustically detectable cellular-level lung injury induced by fluid mechanical stresses in microfluidic airway systems. *Proc. Natl. Acad. Sci. USA* 104: 18886–18891 (2007).

a *lung-on-a-chip* and was one of the first of a class of devices now known as *organs-on-chips*.[30, j]

Organs-on-chips attempt to mimic the physiological and disease mechanisms of human organs, including tumors. By directly testing drugs in small pieces of human tissue, scientists hope to obtain relevant information about diseases and drug efficacy in humans without testing the drugs first in animals. It's an animal-saver technology that scientists hope will be much more cost-effective than the old technology based on animal testing. And for most tissues that are often extracted from the body via a biopsy, as in the case of tumors, the availability of these tissues is scarce, so some level of microfluidic manipulation becomes indispensable. Being small is essential: A microfluidic device can deliver small volumes of many drugs into a small area. Hollywood would like you to believe that scientists can keep whole organs bubbling alive in a glass jar, but we will not be testing drugs like that anytime soon. A large organ must be adequately vascularized and perfused to be kept alive and functioning, whereas a small piece of tissue can be maintained with the help of diffusion—just like skin breathing can sustain small organisms—or clever microfluidic design.[31]

In the first lung-on-a-chip model, the "airways" were microfluidic channels with no particular mechanical function or end structure like the alveoli. Dan Huh decided to head to Harvard's Wyss Institute for a postdoc to continue perfecting his lung-on-a-chip models. The first step was to see if he could culture alveolar epithelial cells (the cells that form the alveoli) in a microfluidic device—which worked. But the real challenge was imitating the dynamic, physiological contraction and expansion of air sacs. "I tried really hard to design some kind of a mechanical stretcher that could be integrated into a microfluidic device where I could culture alveolar cells and then stretch and relax them to mimic breathing motions,"[28] remembers Huh. But none of the designs he could think of would work.

The inspiration for the design came to him one Sunday morning while dozing off at the service of his Protestant church in Brookline. Two side PDMS chambers could be used as pneumatic actuators to pull on the porous membrane seeded with epithelial cells. The chip could replicate mechanical forces coupled with physiological breathing motions. His advisor, Don Ingber, insisted on refining the model by mimicking the alveolar–capillary unit in which alveolar lung cells touch endothelial blood capillary cells. Huh made a sandwich device again, but this time seeded vascular cells underneath. A critical demonstration

---

[j] Around that time, many other groups, such as David Beebe's, Donald Gaver's, Noo Li Jeon's, or my own, to cite a few, had also started developing the concept that human cells in microchannels emulate microphysiological conditions better than open petri dishes.

[30] Ingber, D. E. Human organs-on-chips for disease modelling, drug development and personalized medicine. *Nat. Rev. Genet.* 23: 467–491 (2022).

[31] Horowitz, L. F., Rodriguez, A. D., Ray, T., & Folch, A. Microfluidics for interrogating live intact tissues. *Microsystems Nanoeng.* 6: 69 (2020).

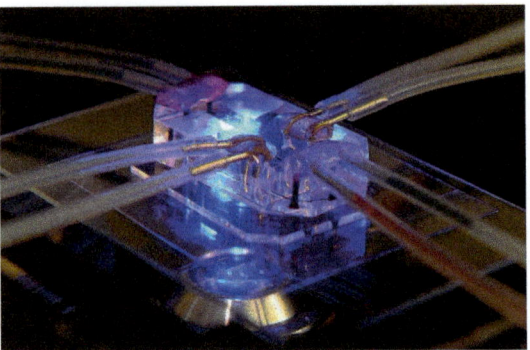

Figure 15.9 Lung-on-a-chip device designed by
Dan Huh.
Courtesy of Wyss Institute at Harvard University.

was the recruitment of immune cells mimicking the natural lung response. Huh
first introduced bacterial cells on the channel containing lung cells, and next,
he placed white blood cells into the channel containing the endothelial cells;
he showed the whole process of immune cell adhesion, with white blood cells
moving through capillary walls and the engulfing of bacteria by a white blood
cell in the lung compartment.[32]

Using the same lung-on-a-chip device (Figure 15.9), Huh was able to model
the process of pulmonary edema induced by drug toxicity observed in cancer patients after treatment with a signaling protein called interleukin-2 (IL-2).
IL-2 is used to stimulate white blood cells into fighting the cancer but can
also be toxic. Without performing a single animal test, this research led to the
identification of new drugs that could prevent IL-2 toxicity in the future.[33]

As a testament to Grotberg's vision, the organ-on-a-chip concept could be
easily extended to other tissues and organs. The pharmaceutical industry immediately took notice.[k] Huh proudly remembers that, right after Ingber and him
published the first paper in *Science*, "we started getting inquiries and calls from
pharma right away."[28] In 2009, Ingber launched Harvard's Wyss Institute, which
now employs more than 300 staff members. Hong Jun Kim also joined the

---

[k] Organs-on-chips have been noticed far beyond the academic and pharma circles. A curator for
the Museum of Modern Art (MoMA) of New York City acquired a replica of Huh's chip that was
exhibited at MoMA for one year during 2015. That same year, the Wyss Institute's human organs-
on-chip won the Design of the Year 2015 Award, the United Kingdom's most prestigious design
award.[28]

[32] Huh, D., Matthews, B. D., Mammoto, A., Montoya-Zavala, M., Hsin, H. Y., & Ingber, D. E.
Reconstituting organ-level lung functions on a chip. *Science* 328: 1662–1668 (2010).

[33] Huh, D., Leslie, D. C., Matthews, B. D., Fraser, J. P., Jurek, S., Hamilton, G. A., et al. A human
disease model of drug toxicity–induced pulmonary edema in a lung-on-a-chip microdevice. *Sci.
Transl. Med.* 4: 159ra147 (2012).

team and, with his background in microbiology, developed the first "gut-on-a-chip"; he used the convenient PDMS stretching capabilities to mimic the gut's peristaltic (waves of squeezing) function.[34] To predict tissue responses to drugs and mimic the physiology of the whole human body, Wyss researchers have developed a platform that links up to ten different organs-on-chips through a standard vascular circuit.

Riding the on-chip revolution conceived by Grotberg and Takayama, organs-on-chips aim to simulate various aspects of human physiology and disease. The Wyss Institute and many others have generated a variety of microphysiological models of living human organs and functions, including the lung, intestine, kidney, skin, bone marrow, the blood–brain barrier, oocyte fertilization, vascular growth, tumor cell migration, and blood clotting, among others. Cancer researchers, including those in my lab, are developing complex tumor-on-a-chip instruments to test cancer drugs more effectively using only small patient biopsies.[35] The patients who donate these biopsies altruistically, knowing very well that the cure will not arrive in time for them, are the real heros here. Thanks to them, scientists can now investigate new treatments with these patient avatars in a way that does not entail the cost, suffering, and ethical issues associated with animal or human testing. Imagine a future when doctors extract a biopsy and, using a tumor-on-a-chip device, they can figure out in less than a week which drug cocktail fights best against your tumor. That is still in the future, but just as Craig Foster was willing to hold his breath for a long time to see if he could learn something from an octopus, I'm eager to hold my breath that the future will be microfluidic.

Summary

- Evolution has come up with three different solutions to the problem of distributing oxygen to all the cells of multicellular animals—and all three solutions are microfluidic: *diffusion* through the skin, the *gill*, and the *lung*. Large, underwater organisms have gills to extract oxygen from water and transfer it into the blood. Large terrestrial organisms—such as all mammals, reptiles, and birds—have lungs that directly transfer oxygen from air into the blood.

---

[34] Kim, H. J., Huh, D., Hamilton, G., & Ingber, D. E. Human gut-on-a-chip inhabited by microbial flora that experiences intestinal peristalsis-like motions and flow. *Lab Chip* 12: 2165–2174 (2012).

[35] Horowitz, L. F., Rodriguez, A. D., Dereli-Korkut, Z., Lin, R., Castro, K., Mikheev, A., et al. Multiplexed drug testing of tumor slices using a microfluidic platform. *Nat. Precis. Oncol.* 4: 12 (2020).

- Not all animals conform to just one of these three types of respiration strictly and combine two or more. For example, lungfish have gills and lungs and can also breathe through their skin.
- Because animal experiments are very poor predictors of outcomes in human disease, lung doctor Jim Grotberg and microfluidic engineer Shuichi Takayama, with their student Dan Huh at the University of Michigan, built a microfluidic chip that creates a living avatar of the lungs' airways. They used this *lung-on-a-chip*, the first *organ-on-a-chip* device, to better understand the lungs' crackling sounds during pneumonia.
- Organs-on-chips aim to more accurately simulate various aspects of human physiology and disease. For example, tumor-on-a-chip devices are being used to better predict cancer drug efficacy and toxicity.

16

The Sounds of Microfluidics

From the Tiny, Fluid-Filled Ancient Circuits of Our Hearing and Balance to the Simplicity of Ultrasound-Powered Chips

Roger Payne was a young assistant professor of biology at Tufts University in the 1960s when he heard on the radio that a whale had washed ashore at nearby Revere Beach in Massachusetts. "Oh great, I'll go see it!" he thought, having never seen a whale in person. But by the time he got there, it was already dark and everyone else was gone. What he saw instead was a mutilated carcass of a dolphin.[a] "Someone had carved their initials on the poor animal. The tail was cut. Someone else had stuck a cigar butt in its blowhole. I was overwhelmingly depressed," he said.[1] At the time, whales and dolphins were not poster animals for the green movement yet (Figure 16.1). Although the days when sperm whales were killed just to make candles from their spermaceti organ were long gone (see Chapter 10), throughout the 1960s more than 60,000 whales were still being exterminated every year with explosive harpoons, only to end up as margarine,[b] animal feed, and fertilizer.[2] For Payne, the shock of that horrendous sight launched him on an entirely different research path than the one he had been on, which had focused on the sounds made by small flying creatures such as bats, owls, and moths.

---

[a] Both dolphins and whales are cetaceans (marine mammals), which explains the broadcaster's confusion.

[b] In 1929, two companies—Lever Brothers, a British soap maker, and Margarine Unie, a Dutch firm—merged to become Unilever. Both companies had discovered how to make margarine from whale oil as its only fat, and they decided to merge instead of competing. Unilever became the world's largest purchaser of oils and fats. By 1935, 84% of the whale oil in the world went into margarine. Unilever bought all the whale oil produced by Norway and sent it to its German factories. Currently, it only makes margarine from plant oils. Now generating 60 billion euros in yearly revenue, Unilever is today one of the world's top companies, the parent company of products ranging from Axe Body Spray to Ben & Jerry's ice cream.[3]

[1] Minoff, A., & Feder, E. Remembering Roger Payne, who helped save the whales. *Science Friday*. https://www.sciencefriday.com/segments/roger-payne-whalesong-recording-undiscovered (2023, June 23).

[2] Rocha, J. R. C., Clapham, P. J., & Ivashchenko, Y. Emptying the oceans: A summary of industrial whaling catches in the 20th century. *Mar. Fish. Rev.* 76: 37–48 (2015).

[3] Laskow, S. Margarine once contained a whole lot more whale. *Atlas Obscura*. https://www.atlasobscura.com/articles/what-is-margarine-made-of (2018, October 11).

Figure 16.1 (Top) A mother sperm whale and her calf. Humpback whales often sing chorus songs in feeding grounds and migration routes. (Bottom) A humpback whale in the process of breaching.

Image sources: (Top) User Gabriel Barathieu on Wikipedia. https://en.wikipedia.org/wiki/ Sperm_whale#/media/File:Mother_and_baby_sperm_whale.jpg (CC BY-SA 2.0). (Bottom) User Whit Welles on Wikipedia. https://en.wikipedia.org/wiki/Humpback_whale#/media/ File:Humpback_stellwagen_edit.jpg (CC BY 3.0).

In the 50s and 60s, at the peak of the Cold War, the U.S. Navy led extensive studies of underwater sound. The Navy wanted to distinguish the sonar's blip echoing off a Russian submarine from other sounds, such as a moaning whale and other creatures, that the sonar's hydrophone (an underwater microphone) was picking up as background noise. Payne, an animal sound expert, was invited to visit Navy engineer Frank Watlington in Bermuda in 1967 to listen to whale sound recordings that the Navy had collected.[4] "The sounds that I heard were absolutely transforming. They were shocking. I had never heard

[4] Van Dine, A. Remembering Roger Payne, whose recordings of whale songs changed the world. *Vermont Public.* https://www.vermontpublic.org/local-news/2023-06-14/remembering-roger-payne-whose-recordings-of-whale-songs-changed-the-world (2023, June 14).

any animal make any sound even approximately as intriguing and command-ing. It was incredible," he said.[1] Payne, a pretty serious amateur cellist, had heard *music*.

When Payne recalled the whale sounds he had heard as he stood in front of the dolphin carcass, it dawned upon him: If he could get humanity to listen to these moans and understand them as *songs*—sounds repeated in complex patterns—he could get humans interested in whales enough to save them. Animal music, Payne presumed, would inspire empathy between people and animals just as human music does between people. "Music, and the appreciation thereof, is older than we are by millions of years," he said.[5]

Payne decided to go on television and radio shows to promote the notion that we should not make butter and fertilizer out of animals that sing. He also published a groundbreaking acoustic analysis of Watlington's recordings, prov-ing that these were, indeed, songs.[6] In 1970, Payne produced the celebrated multiplatinum album *Songs of the Humpback Whales*,[c] used in NASA's Golden Record aboard the two 1977 Voyager spacecraft,[d] and re-recorded by singers and orchestras who played with the whale choir in the background.[7] It struck the chord he was looking for. In 1972, Congress passed the Marine Mammal Protection Act, and ten years later, commercial whaling was officially banned in the United States.[3] Payne is now widely recognized as the man who inspired the public outrage and the Greenpeace movement that ultimately saved the endan-gered whales.[1] To the delight of human audiences, his album stimulated further research[e] into an underwater realm of whale pods that sing arias, duets, and choruses across thousands of miles inside an ocean that previously sounded practically silent to us.

Whales and all vertebrate animals vocalize sounds *to be heard*, and hearing occurs through a microfluidic biological sensor. Some animals make a richer

[c] Payne's album, which included the recordings by Frank Watlington and also others taken by Payne and his wife Katy, quickly sold more than 100,000 copies and eventually went multiplatinum (i.e., more than 2 million copies sold). Artists who have used excerpts from Payne's album are numer-ous, including Judy Collins, Pete Seeger, Kate Bush, Glass Wave, as well as Alan Hovhaness in his symphonic suite *And God Created Great Whales*. It was also used in the movie *Star Trek IV: The Voyage Home*. In 2010, the National Recording Registry inducted the album into one of the signifi-cant recordings that "are culturally, historically, or aesthetically important, and/or inform or reflect life in the United States."[7]

[d] Both Voyagers, still carrying the whale sounds, are now flying beyond the Solar System in interstellar space.

[e] Payne founded Ocean Alliance (https://www.whale.org), one of the major funding agencies for whale research.

[5] Folds, B. Roger Payne—Creativity is the only way to save humanity: Lightning bugs w/ Ben Folds. YouTube. https://www.youtube.com/watch?v=vkhWFmwcBM8 (2022).

[6] Payne, R. S., & McVay, S. Songs of humpback whales. *Science* 173: 585–597 (1971).

[7] Ocean Alliance. Songs of the humpback whale. https://whale.org/humpback-song (2024).

variety and reach a lower or higher frequency of sounds than others.[f] Cows mostly moo and bellow but also snort and grunt. Cats meow, yowl, purr, trill, and hiss. Dogs bark when excited; growl and huff when they feel threatened; whine, whimper, or howl when sad; and yelp when injured. Birds can chirp, whistle, trill, croak, sing, and more. As Payne taught us, humpback whales not only sing but also speak different dialects and learn new tunes from each other on their annual migrations. Some animals, such as dolphins and bats, produce sound for locating prey or obstacles based on the echo of that sound, an ability called *echolocation*. Mice chirp and squeak—when we hear them—but most mouse communication is produced in ultrasonic vocalizations beyond human hearing like bats do.[8] Gibbons, a type of lesser ape, croon unique melodies—often in duets and with "accents" varying across their Southeast Asian rainforest habitat—that they use to call members of their family unit or to send an alert if an intruder has entered their space. The males of many frog species worldwide gather to croak out in a chorus when they want to attract a female.[9] And humans smack kisses, talk, whisper, whine, scream, shriek, groan, moan, burp, laugh, giggle, hum, whistle, clap, snap their fingers, play instruments, and chant alone, in duets, and in choruses, too. These are just the sounds we humans recognize, but as with whales and other species, we will hear more if we listen carefully.

None of these animals would produce these sounds were it not for the fact that they are meant to be heard. Despite the variety of sounds, the sensing mechanism is very similar in all vertebrates due to a common ancient organ that appeared hundreds of millions of years ago in the evolution tree.[10] And, perhaps because animal movement is inextricably linked to survival, the inner ear houses two evolutionarily related sensory organs: one for sound that detects frequency and volume, and the other for movement that detects balance and acceleration. All vertebrates—from a mouse[12] to a whale[13]—have very similar, tiny motion and sound sensors in their inner ear; without exception, they consist of microfluidic

[f] Some fish hear well and some do not, using a similar but more primitive microfluidic hearing organ. Many animals—from elephants to ants and snakes—sense vibrations through their bodies. Some insects that make sounds—grasshoppers, crickets, cicadas, and moths—have hair-like structures located on their legs, abdomens, or antennae directly exposed to air and are therefore not microfluidic.[11]

[8] Vogel, A. P., Tsanas, A., & Scattoni, M. L. Quantifying ultrasonic mouse vocalizations using acoustic analysis in a supervised statistical machine learning framework. *Sci. Rep.* 9: 8100 (2019).

[9] Olivia—F&F Team. In the wild, which animals can sing? Flora & Fauna. https://www.floraandfauna.com.au/blog/in-the-wild-which-animals-can-sing (2022).

[10] Wever, E. G. Sound reception in vertebrates—Auditory mechanisms of fishes and amphibians. *Encyclopedia Britannica*. https://www.britannica.com/science/sound-reception/Sound-reception-in-vertebrates-auditory-mechanisms-of-fishes-and-amphibians (2022).

[11] Wever, E. G. Organs of sound reception in invertebrates. *Encyclopedia Britannica*. https://www.britannica.com/science/sound-reception/Organs-of-sound-reception-in-invertebrates (2022).

[12] Yin, H. X., Zhang, P., Wang, Z., Liu, Y. F., Liu, Y., Xiao, T. Q., et al. Investigation of inner ear anatomy in mouse using X-ray phase contrast tomography. *Microsc. Res. Tech.* 82: 953–960 (2019).

[13] Ketten, D. Blue whale inner ear. *EurekAlert!* https://www.eurekalert.org/multimedia/537165 (2023).

channels (Figure 16.2, left). So it is entirely possible that some animals might have already been singing for millions of years, long before humans set foot on Earth, and perhaps even long before whales began singing the songs revealed to us by Payne.

What is sound, and how is it transformed into the electrical signals sent to your brain? Sound is an audible pressure wave. When it reaches your ear through the air, it travels for about an inch through the ear canal until the fluctuating pressure hits a membrane called the *eardrum*, pushing it back and forth and causing it to vibrate.[g] On the other side of the eardrum, a trio of tiny bones[h] transmits the vibrations to another, twenty-times-smaller membrane called the *oval window*. This mechanical relay acts as a twenty-fold pressure amplifier to mobilize the fluid on the other side of the oval window.

The oval window is the entrance to our *cochlea*—named for its coil like a snail shell[i]—the microfluidic microphone of our inner ear.[j] If you were a nanorobot entering the cochlea at its *oval window*, you would swim up the spiral through a microfluidic channel called the *vestibular duct*. Upon reaching the top of the cochlea—like in a modern parking garage—you would go down through a different microfluidic channel parallel to the vestibular duct called the *tympanic duct*. The fluids do not go in and out, so the tympanic duct ends in a membrane called the *round window* next to the oval window that accommodates the fluid displacements caused by the input sound's pressure. However, the most essential part of the cochlea (visualized as a straight channel for clarity in Figure 16.2, right) is a third and narrower microfluidic channel. This *cochlear duct* (shown in cyan in Figure 16.2) is also coiled and sandwiched between the vestibular and tympanic ducts. In the cochlear duct resides the *organ of Corti*, a tissue structure containing sound-sensitive nerve cells called *hair cells*.[k]

The organ of Corti is the body's microphone. The hair cells receive their name because they have a punk hairdo, with tiny hair structures called *stereocilia* sticking out in bundles (Figure 16.3). Crucially, the spiraling parking garage is not

---

[g] Many aquatic mammals have adapted their inner ear so that sound passes readily through the middle ear into the inner ear. In whales, for example, the eardrum serves no useful purpose.[14]

[h] These tiny bones or *ossicles* were given their Latin names for their distinctive shapes: *malleus* (hammer), *incus* (anvil), and *stapes* (stirrup).

[i] "Cochlea" is borrowed from Latin ("a snail") and derives from Ancient Greek κόχλιας (kókhlias, "a snail with a spiral shell").

[j] The name cochlea also applies to species in which the organ is not coiled, as in reptiles, birds, and egg-laying mammals. In some mammalian species, the coil of this spiral turns as little as twice, and in others it turns as many as four times.[15]

[k] We are born with approximately 16,000 hair cells, but when they are exposed to loud noises very often, they die (the high-frequency-sensitive cells first) and cannot regenerate. The loss of hair cells makes it harder for a person to hear conversations.

[14] Wever, E. G. Hearing in subhuman mammals. *Encyclopedia Britannica*. https://www. britannica.com/science/sound-reception/Hearing-in-subhuman-mammals (2022).

[11] Wever, E. G. Organs of sound reception in invertebrates. *Encyclopedia Britannica*. https://www. britannica.com/science/sound-reception/Organs-of-sound-reception-in-invertebrates (2022).

[15] Wever, E. G. Hearing in mammals. *Encyclopedia Britannica*. https://www.britannica.com/ science/sound-reception/Hearing-in-birds#ref64820 (2022).

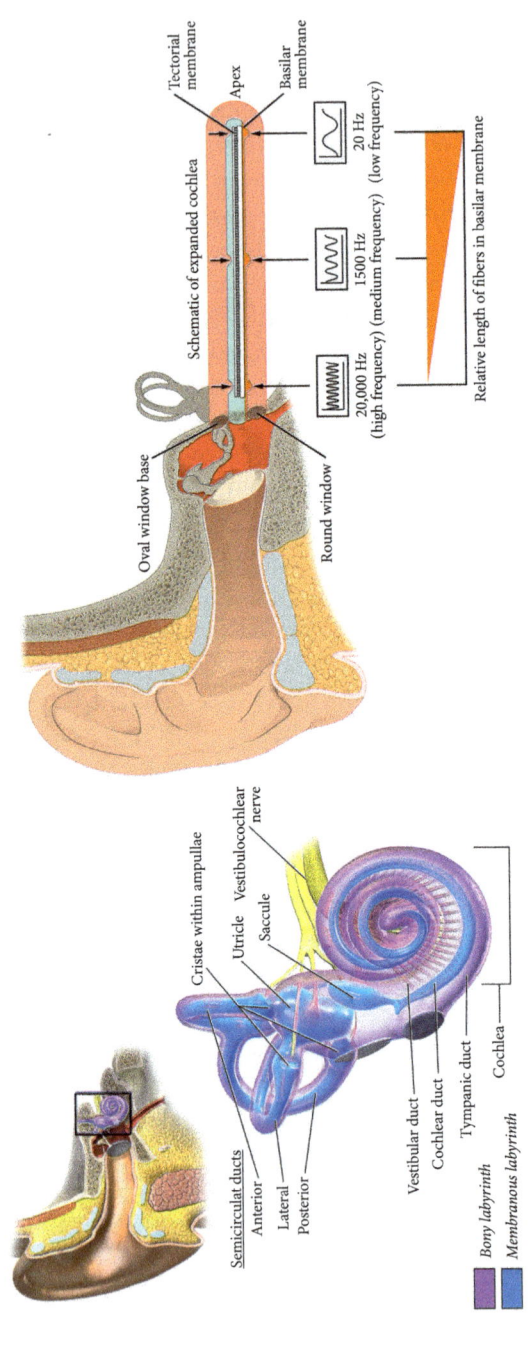

Figure 16.2 (Left) The human inner ear. The inner ear of a mouse and that of a whale are, surprisingly, very similar in overall structure to ours. (Right) Schematic representation of the cochlea as a straight channel, depicting its frequency coding capabilities.

Image sources: (Left) Blausen.com staff. Medical gallery of Blausen Medical 2014. Wikipedia. https://en.wikipedia.org/wiki/Inner_ear#/media/ File:Blausen_0329_EarAnatomy_InternalEar.png (CC BY 3.0). (Right) Textbook OpenStax Anatomy and Physiology. Wikipedia. https://en.wikipedia.org/wiki/ Inner_ear#/media/File:1408_Frequency_Coding_in_The_Cochlea.jpg (CC BY 4.0).

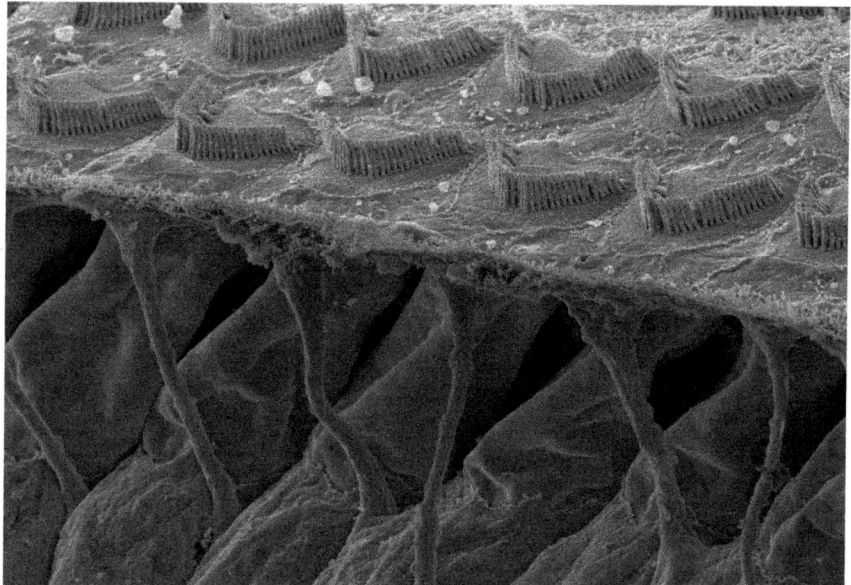

Figure 16.3 Scanning electron micrograph of the organ of Corti, with the hair cells' stereocilia hair bundles sticking out at the basilar membrane surface where they are normally exposed to vibrating fluid. Each "V" shape corresponds to one hair cell and is approximately five thousandths of a millimeter wide.
Image courtesy of Prof. David Furness, Keele University.

built with rigid cement; on the contrary, the walls that separate the cochlear duct from the vestibular and tympanic ducts are both flexible membranes. This flexibility is essential because the hair cells of the organ of Corti are attached atop the *basilar membrane,* which separates the cochlear duct from the tympanic duct.

The auditory system responds quickly because all the parts are small and submerged in a tiny fluid volume. The cochlea of a modern human is only 1 centimeter wide and the 2.5-turn spiral path is roughly 3 centimeters long.[16] When a sound wave knocks at the cochlea's oval window, it sends vibrations through the fluid down the vestibular duct. These vibrations cross the membrane that separates the cochlear duct and finally cross the basilar membrane into the tympanic duct to die off at the round window. With each pulse of the basilar membrane and its hair cells, the undulating fluid causes the hair bundles to bend back—much like when your hair flows with an incoming wave when you

---

[16] Jagt, A. M. A. V. Der, Kalkman, R. K., Briaire, J. J., Verbist, B. M., & Frijns, J. H. M. Variations in cochlear duct shape revealed on clinical CT images with an automatic tracing method. *Sci. Rep.* 7: 17566 (2017).

swim at the beach. The hair bundles and the basilar membrane in the human ear can respond to up to 20,000 beats per second. The bending of the hair bundles triggers a nervous impulse in the hair cells, which signals to our brain that sound has been detected. Each species has a different auditory system geometry, which determines the range of frequencies it can hear.[1]

The shape of the basilar membrane is crucial for our ability to discriminate high and low pitches and, thus, to appreciate music. The basilar membrane is narrower toward the entrance of the cochlea, progressively widening up to five times toward the end of the cochlea (even in species with an uncoiled cochlea, like birds and reptiles).[15] Remember how, in a band, a large drum produces a lower pitched sound than a small drum? The basilar membrane resonates with different frequencies in a similar manner. When a sound wave enters the cochlea, a high-pitched sound such as the chirp of a bird will cause the basilar membrane to vibrate toward the cochlea's entrance—where the membrane is narrowest—stimulating only those hair cells (see schematic of the uncoiled cochlea in Figure 16.2, right). On the other hand, a low-pitched sound like a thunder will cause the basilar membrane to vibrate toward the apex of the spiral—where the basilar membrane is widest—stimulating a different set of hair cells. As a result, your hair cells transform the frequency of sounds into a spatial distribution of nerve impulses (see Figure 16.2, right).[m] The brain then interprets this spatial map as the frequencies of the sounds.[n]

[1] Sound is a pressure wave or, in other words, a pressure oscillation that travels in space. The frequency of a sound wave is the number of pressure oscillations that pass a given point every second. A high-pitched sound like the one produced by a small flute consists of a lot of short oscillations moving out of the flute every second, whereas the sound of a large trombone consists of fewer, slower fluctuations resulting in a lower pitch. Sound frequency is measured in cycles per second or *hertz*. We can hear between 20 and 20,000 hertz.

[m] There is a plethora of educational videos that illustrate the structure and function of the inner ear with enlightening three-dimensional (3D) graphics and explanations. I particularly enjoyed the 3D animations of auditory transduction and of the cochlea's overall structure by Brandon Pletsch (National Science Foundation award-winning film)[17] or by CrashCourse,[18] of the basilar membrane by The Science Tutorials Channel,[19] and of the organ of Corti by animacionesplus.[20] For the vestibular system, the reader is referred to the videos by Alila Medical Media[21] and by 3D Anatomy Lyon.[22]

[n] Sounds contain many superimposed frequencies called *harmonics*. For example, the same notes in a piano and a trumpet contain different harmonics. The basilar membrane detects all the harmonics of a given sound simultaneously by vibrating at various locations at once.

[17] Pletsch, B. Auditory transduction. YouTube. https://www.youtube.com/watch?v=46aNGGNPm7s (2009).

[18] CrashCourse. Hearing & balance: Crash course anatomy & physiology #17. YouTube. https://www.youtube.com/watch?v=Ie2j7GpC4JU (2015).

[19] The Science Tutorials Channel. Basilar membrane and its response to sound animation: How we hear different pitches. YouTube. https://www.youtube.com/watch?v=vhizzJApNV0 (2021).

[20] Animacionesplus. Ear organ of Corti (full version). YouTube. https://www.youtube.com/watch?v=1JE8WduJKV4 (2010).

[21] Alila Medical Media. The vestibular system, animation. YouTube. https://www.youtube.com/watch?v=ryGMI3SpxCE (2022).

[22] 3D Anatomy Lyon. Vestibular proprioception. YouTube. https://www.youtube.com/watch?v=ZiFyIfBWyOo (2015).

We also owe microfluidics our ability to detect acceleration, head orientation, and balance. However, some people have a better sense of orientation, position, and balance than others. When I was a kid and tried to flip from the swimming pool's diving board, I would lose my sense of up and down in mid-air and invariably back-flopped with a big splash. On the other extreme, we have Olympic gymnast Simone Biles—possibly the best gymnast in history.[o] She can do cartwheels and somersaults on the 10-centimeter-wide balance beam with the ease and gracefulness of a cat (Figure 16.4). Once, in the 2019 U.S. Gymnastics Championship, when she landed on the floor with a radiant smile after a never-seen-before double flip and a double twist, the commentator exclaimed, "So many people cannot even perform this skill on their floor exercise!" Likewise, Johnny Moseley, the influential mogul skier, can keep track of his orientation with supernatural ease. He invented a seemingly impossible off-axis trick in the 1999 Winter X Games: rolling his body three times over in the air while crossing the skis—the *dinner roll*. "He mixes the water, the little flour, as the yeast breeds, the oven bakes it, the dough rises, and he delivers a fresh dinner roll!" screamed the broadcaster excitedly. In a feat of balance, Scottish street trials and mountain bike rider Danny MacAskill has done on his bicycle what most people cannot do on their feet. He has somersaulted with his whole bike over dry-stack walls; ridden his bike on a suspended chain between two parking poles; and climbed, turned, or flipped on either the front or the back wheel on top of bars thinner than a person's arm with the delicacy of a ballet dancer. These athletes achieved what appears otherworldly to most of us with the help of an exceptional sense of balance and orientation.

Our body uses a microfluidic biosensor to detect head position and acceleration using hair cells similar to the ones we use for hearing. The *semicircular ducts* or *canals*, right next to the cochlea (see Figure 16.2, left), consist of three tiny, 6.5-millimeter-diameter fluid-filled ring-shaped tubes at right angles.[23] The orientations of the semicircular canals cause head movement in different planes to stimulate different canals. Each semicircular canal ends in a widening where the sensory organ of rotation, the *crista ampullaris* (or ampullary crest), resides.[p] This poetically named structure consists of forty to seventy hair cells covered by a gelatinous blob. When the head rotates, the fluid above the crista ampullaris lags behind due to inertia, causing the blob to bend along with the

---

[o] Biles has won nineteen gold medals at the world championships, more than any men's or women's gymnast in history. She became the youngest person to ever receive the Presidential Medal of Freedom at the age of twenty-five years.

[p] Thus, the brain receives input from three pairs of crista ampullaris in total, counting both ears.

[23] Daocai, W., Qing, W., Ximing, W., Jingzhen, H., Cheng, L., & Xiangxing, M. Size of the semicircular canals measured by multidetector computed tomography in different age groups. *J. Comput. Assist. Tomogr.* 38: 196–199 (2014).

Figure 16.4 Simone Biles performing the balance beam exercise in the 2016 Olympic Games in Rio de Janeiro, Brazil.

Image source: User Fernando Frazão/Agência Brasil on Wikipedia. https://en.wikipedia.org/wiki/ Simone_Biles#/media/ File:Simone_Biles,_na_prova_final_da_trave_nos_Jogos_Ol%C3%ADmpicos_Rio_2016.jpg (CC BY 2.0).

hair cells' stereocilia. Any bending of these stereocilia in the three semicircular ducts tells the brain that your head has rotated around the corresponding axis.[21]

Two additional microfluidic chambers attached to the semicircular canals detect *linear accelerations* (or decelerations), such as the thrust we experiment when we press the gas pedal (or the brakes) in a car, or the effect of the *gravitational force*, whether standing or lying down.[22] Their names—the *utricle* ("leather bag" in Latin) and the *saccule* ("pouch")—don't quite exude the elegance of "crista ampullaris." The utricle perceives positions and linear accelerations in the horizontal plane. In contrast, the saccule does it in the vertical plane (so it is sensitive to gravity). The utricle and saccule are also referred to as the *otolithic organs* because they both contain a patch of 5,000 hair cells (similar to the cells found in the cochlea) embedded in a gelatinous *otolithic membrane*, so called because the membrane is weighted down with tiny calcareous pebbles or "ear rocks."[q]

These ear rocks add to the inertia of the membrane when there is movement to enhance the sense of motion and gravity. Start running, and your utricle's ear rocks shift backward due to inertia with respect to the hair cells under the

[q] From the Greek *oto* = ear and *lithos* = rock.

otolithic membrane.[r] Lie down, and your ear rocks pull down on your saccule's otolithic membrane due to gravity. The shifts of the pebbles cause the bending of the hair cells' stereocilia, which send the signal to the brain to inform it that you are accelerating or lying down. With the help of additional input from the eyes and the muscles, the brain integrates the information it receives from these ear rock patches swishing back and forth inside a microfluidic chamber and computes how the body is accelerating forward or backward, left or right, or up or down.[21,22] Simply put, if inside each of your ears you have the equivalent of a microfluidic microphone to detect sound, you also have a 3-axis microfluidic accelerometer next to it.

Our ability to hear a wide variety of sounds and perceive our own motion and head position is a gift we owe to evolution. The first proto-fishes swam in the ocean in the Cambrian Period more than 500 million years ago. They were the ancestors of all vertebrates, including dinosaurs and present-day fish, frogs, snakes, lizards, crocodiles, birds, and mammals. Their head had two eyes and no jaws. We infer they must have had something similar to what all their descendants inherited: a vestibular system with three rings at right angles and a crista ampullaris each, as well as the two otolithic organs (the utricle and saccule).[s,10]

Can we tell whether these primitive fish heard sounds? Fish do not have a cochlea, yet many fish can hear through their ear rocks because an underwater sound wave is just a pressure disturbance in the same medium where the fish is swimming.[24] Fish have also developed a related accessory sensory system running from head to tail called the *lateral line*. It is made of a tiny channel with hair cells and small pores, allowing fish to detect weak water motions and pressure gradients along their body length.[25,26] It's as if we had developed a very long, uncoiled second cochlea that reached from our head down to our feet to help us hear better. It is possible that fish first evolved the sensors that allowed them to control their motion and then developed the neural circuits to interpret those water vibrations as underwater sounds. Fish use their vibration-sensing microfluidic organs to localize predators and prey, avoid obstacles, communicate, and orient themselves in laminar and turbulent flows.[25,26] Because large animals necessarily produce larger pressure disturbances during swimming, a

---

[r] The hair cells fire nervous impulses at a frequency that reaches a maximum and a minimum when the stereocilia bend in the direction or away from movement, respectively, but any direction in between (*e.g.*, a lateral acceleration) produces intermediate firing frequencies.[22]

[s] The only two known exceptions are two eel-looking fish: lampreys, which have two semicircular canals, and hagfish, with a single semicircular canal. Birds also have a second vestibular organ, called the lumbosacral canal, that appears to be involved in stabilizing the body during walking.

[24] Ladich, F., & Schulz-Mirbach, T. Diversity in fish auditory systems: One of the riddles of sensory biology. *Front. Ecol. Evol.* 4: 174758 (2016).

[25] Bleckmann, H., & Zelick, R. Lateral line system of fish. *Integr. Zool.* 4: 13–25 (2009).

[26] Braun, C. B., & Coombs, S. The overlapping roles of the inner ear and lateral line: The active space of dipole source detection. *Philos. Trans. R. Soc. B Biol. Sci.* 355: 1115–1119 (2000).

shark stealthily approaching small fish from behind may sound to them like a rumbling freight train sounds to us.

Of course, no animal needed to detect sound in the air before the first vertebrates decided to venture onto land about 400–375 million years ago.[27] Paleontologists have proposed that hair cells appeared around 300 million years ago to help vertebrates such as lungfish and salamanders catch buzzing insects,[28] or perhaps as an extra sensory input for animals living in the dark. Indeed, owls, cats, and geckos that are well adapted to dimly lit environments also have particularly sharp ears.[29] Fossils reveal that nearly 200 million years ago, well before the appearance of whales 50 million years ago, early mammals had very short (2-millimeter-long), uncoiled cochleas that could probably not discern as broad a range of frequencies as present mammals. The coiled cochlea evolved approximately 120 million years ago, likely a space-saving solution to endow mammals with longer basilar membranes and, thus, a broader range of hearing frequencies.[30]

All these hypotheses are difficult to prove because these mechanisms evolved very deep in the tunnel of time. Still, one thing is clear: Nature never abandoned the common motif of transducing sound via hair cells in microfluidic channels in all its design iterations of the various hearing systems in vertebrates. As far as we know, *all* sounds *ever* detected by *any* vertebrate animal on Earth have been heard with a microfluidic organ for hundreds of millions of years.

* * *

Scientists have known for a long time that sound is a pressure-density wave that allows us to communicate and enjoy music, but it can also be used to push things around. Density waves have familiar manifestations: Leaves and twigs floating on a swimming pool accumulate on one side when children produce ripples on the other side of the pool's surface. Like coffee and milk, mixing molecules and cells is always easy—it's *separating* them that takes effort. This tendency of systems to become disorganized—called *entropy*—seems to be a fundamental property of our Universe, and there is nothing we can do to fight it. The same thing happens with decks of cards. In a few seconds, you can mix a deck of red cards with a deck of blue cards, but it takes a lot more time and energy to organize the decks into two separate packs again. That's why researchers got very excited when they found that under certain conditions, blasting a fluid

[27] Pennisi, E. Genes for life on land evolved earlier in fish: Hidden genetic pathways behind land–water transition are found in living fish. *Science* 371: 658–659 (2021).

[28] Müller, J., & Tsuji, L. A. Impedance-matching hearing in Paleozoic reptiles: Evidence of advanced sensory perception at an early stage of amniote evolution. *PLoS One* 2: e889 (2007).

[29] Balter, M. Let's hear it for the first ears. *Science.* https://www.science.org/content/article/lets-hear-it-first-ears (2007).

[30] Manley, G. A. Evolutionary paths to mammalian cochleae. *J. Assoc. Res. Otolaryngol.* 13: 733–743 (2012).

mixture with high-frequency sound, or *ultrasound*, could separate it into its components—a technique called acoustophoresis.

Acoustophoresis means "migration by sound." Researchers place an ultrasound buzzer in contact with the exterior wall of the channel. When the buzzer is active, the wall vibrates and transmits that vibration across the width of the channel. A density wave is stabilized if the vibration period (which can easily be tuned) is an exact multiple of the time it takes for the wave to bounce back from the opposite wall of the channel. Riding these so-called standing waves, molecules, particles, and cells floating through the channels reorganize according to their density and size (Figure 16.5).

Acoustophoresis is a phenomenon that has been documented for several decades. The first acoustophoresis experiments done in the early 1990s consisted of observations that the application of ultrasonic standing waves to blood vessels caused the segregation of the red blood cells into bands.[31,32] Ultrasound-based

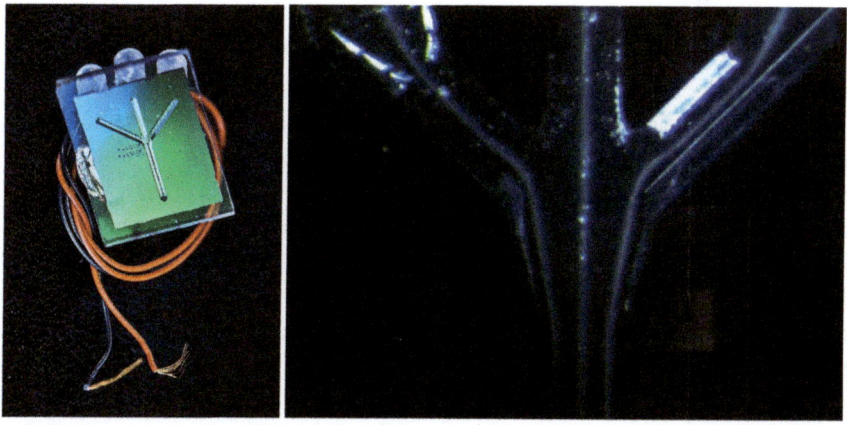

Figure 16.5 Researchers use acoustophoresis to separate particles in a mixture, such as blood cell types in blood, by the simple application of ultrasound waves in a microchannel. (Left) An assembled device (with the ultrasound transducer underneath). (Right) A device in operation, with the white traces of fluorescent particles separated by laminar streams in a channel 375 micrometers wide. Flow comes from below.

Sources: (Left) Courtesy of CBMS (Okaar Photography). (Right) Courtesy of Laurell lab.

[31] Dyson, M., Woodward, B., & Pond, J. B. Flow of red blood cells stopped by ultrasound defence and territorial behaviour dissociated by hypothalamic lesions in the rat. *Nature* 232: 572–573 (1971).
[32] Baker, N. V. Segregation and sedimentation of red blood cells in ultrasonic standing waves. *Nature* 239: 398–399 (1972).

separation in microfluidic channels was demonstrated around 2000.[33-35] For the first time, acoustophoresis had made separation almost as easy as mixing. It's as if ultrasound had magically segregated the red cards from the blue cards into distinct piles.

Before acoustophoresis, mixing had been easy, but separation had been challenging, requiring techniques with complex instrumentation. For example, separation by centrifugation uses a bulky and expensive centrifuge, not unlike the one inside your washing machine. Separation through electrical fields—*electrophoresis*—involves a voltage generator as big as your microwave capable of generating large voltages. And separation using magnetic fields—*magnetophoresis*—entails large magnets and some means of positioning the device or the magnets to activate and deactivate the magnetic field by proximity.

By contrast, acoustophoresis only necessitates an ultrasound buzzer that costs a few dollars and can be computer-controlled—or better yet, battery-operated—because it is a low-voltage device. For researchers interested in diagnostics, such as the Swede Thomas Laurell, the development of acoustophoresis in the early 2000s inspired new avenues of research.

Laurell grew up in Lund, went to school in Lund, and became a professor in Lund, a quaint city with 90,000 inhabitants and full of cobblestoned streets at the southern tip of Sweden. He recalls colleagues at his department using ultrasound to create streaming in fluids; they studied the viscous properties of milk inside large closed containers to determine freshness. At that time, he got excited by that study because in his work on chemical reactions in microchannels, he had observed that laminar flow impeded mixing. Thus, he assumed that faster mixing by ultrasound-induced streaming would accelerate the chemical reaction. Unaware that the streaming mechanism was a macroscale effect and would not work in microchannels, he went on to test this idea further. In 1999, he introduced tiny beads into one of the inputs of the channel to determine if he could see the beads swirling around when he turned the ultrasound on. To his surprise and disappointment, the beads organized in bands instead. Initially, he thought it was a failed experiment.

---

[33] Hill, M., & Wood, R. J. K. Modelling in the design of a flow-through ultrasonic separator. *Ultrasonics* 38: 662–665 (2000).
[34] Harris, N. R., Hill, M., Beeby, S., Shen, Y., White, N. M., Hawkes, J. J., & Coakley, W. T. A silicon microfluidic ultrasonic separator. *Sens. Actuators B Chem.* 95: 425–434 (2003).
[35] Hawkes, J. J., & Coakley, W. T. Force field particle filter, combining ultrasound standing waves and laminar flow. *Sens. Actuator B Chem.* 75: 213–222 (2001).

"At first, I didn't fully understand it," he recalls. This type of honesty is unusual in gray-haired professors who often boast about their wisdom, but he just smiles with a serenity that transports you to his small town.

A few months later, a thoracic surgeon named Henrik Jönsson contacted him. He had seen experiments in which ultrasound applied to a container full of blood caused the blood cells to separate in bands and sediment.[31,32] Jönsson wanted to use this principle to separate blood cells and isolate plasma for his patients. That's when Laurell recalled his failed experiment and realized that his bead experiment could be repeated with blood cells to separate the cells from the plasma faster—and more precisely—if it was done in a microchannel. He was correct.[36–38] Scores of researchers developing various applications have used their technique. Although it is not in the hands of doctors yet, Laurell and colleagues have envisioned a future where large centrifuge apparatuses in hospitals and clinical research labs will be replaced by much simpler, sound-operated microfluidic tools.[39]

* * *

With its inherent beauty, microfluidics has also spilled over into other realms of knowledge. Linden Gledhill is an artist, although his setup is reminiscent of acoustofluidics. He has developed a set of unique tools to *capture* on film fluids that are set in motion using sound. His art grew from the time he spent as a kid sharing his dad's favorite hobbies. "My dad was an engineer, mechanical-maintenance kind of guy, so we built things together. But he was also interested in photography, and I picked up the hobby with him," said Gledhill, who also remembers, at the age of eleven or twelve, building a telescope and an infrared trigger from scratch with his dad.[40] Now Florida-based, he is a biochemist-by-day-turned-artist-by-night who likes to freeze-photograph all sorts of small things in motion in his garage—from ferrofluids that become uplifted when they flow near a magnetic field to agitated wasps in flight and, dazzlingly, the ripples and ejecta caused by sound

[36] Nilsson, A., Petersson, F., Jönsson, H., & Laurell, T. Acoustic control of suspended particles in micro fluidic chips. *Lab Chip* 4: 131–135 (2004).

[37] Petersson, F., Nilsson, A., Holm, C., Jönsson, H., & Laurell, T. Separation of lipids from blood utilizing ultrasonic standing waves in microfluidic channels. *Analyst* 129: 938–943 (2004).

[38] Lenshof, A., Ahmad-Tajudin, A., Järås, K., Swärd-Nilsson, A. M., Åberg, L., Marko-Varga, G., et al. Acoustic whole blood plasmapheresis chip for prostate specific antigen microarray diagnostics. *Anal. Chem.* 81: 6030–6037 (2009).

[39] Wu, M., Ozcelik, A., Rufo, J., Wang, Z., Fang, R., & Jun Huang, T. Acoustofluidic separation of cells and particles. *Microsystems Nanoeng.* 5: 32 (2019).

[40] Mosher, D. Cheap DIY camera systems perform amazing photographic feats. *Wired.* https://www.wired.com/2010/12/gledhill-cognisys-photos/?pid=741&viewall=true (2010).

waves on the surface of inks.[41] The method he uses to set fluids into motion with sound is reminiscent of the old ultra-high-speed photography technique developed by Belgian photographer Frans Vandemaele[t] that consists of taking pictures of drops falling and splashing on the surface of liquids, with the difference being that the drops are not falling—they are launched into the air from below.[42]

In 2011, Canon released a new set of toner dyes for its BubbleJet printers, and the Canon engineers asked Gledhill if he would be willing to shoot a commercial that captured the vibrant color of the inks.[43] Gledhill's setup resembles a dynamic Jackson Pollock three-dimensional canvas in miniature. He deposits drops of ink atop the small acoustic membrane from a stereo speaker. Then, he vibrates the membrane with particular notes to *see* their effect on the ink. The vibration causes the drops to dance atop the membrane. In ultra-slow motion, the drops reveal mixing patterns of arresting beauty (Figure 16.6).[u] The pattern is different for each note, every time—a sort of unrepeatable microfluidic ballet sculpture we feel fortunate to witness through Gledhill's camera. Like a limitless combination of Pollock paintings in the making, the images freeze for the observer the moment before the final splash where the paint drips mix and meet the canvas.

These images of high-speed paint dripping are not just strikingly gorgeous. They also evoke how surface tension makes little water sculptures around us daily—when a faucet drips if we forget to close it entirely, or when a wet rag trickles as we wring it out. The rebounds captured in Gledhill's snapshots remind me of the smell of sea spray, the excitement of children stomping on rain puddles,

[t] Gledhill is the first to acknowledge Vandemaele: "Frans Vandemaele developed the method and did the original work, which inspired me to try it," he told me. Vandemaele is better known as *fotoopa* on Flickr. See https://www.flickr.com/photos/fotoopa_hs/collections/72157618822999514. Vandemaele is now past eighty but happy that his technique is being used by so many photographers throughout the world. A Flickr "pool" containing more than 1,400 similar images can be found at https://www.flickr.com/groups/1496151@N25/pool. Note that this pool mixes high-speed images of falling drops and those generated by fotoopa's sound-based technique.

[u] The process is explained in detail by Lyon.[43] Gledhill's full "Water figures" gallery with more than 200 photos is available at https://www.flickr.com/photos/13084997@N03/sets/72157608258335431. The movies can be seen at https://www.flickr.com/photos/dentsulondon/sets/72157624828897635.

[41] Mosher, D. Trippy photos show how beautiful water can look when it's blasted with sound. *Insider.* https://www.businessinsider.com/water-slow-motion-video-standing-waves-cymatics-2018-5?fbclid=IwAR242kvo7RQL9O4xo_XpOCLfl5-L_-nbnte_I1t74HVd6JD-BBcoHXaFJhg (2018, May 29).

[42] O'Neill, C. How sound can create sculpture. NPR. https://www.npr.org/sections/pictureshow/2011/04/13/135377949/how-sound-can-create-sculpture (2011, April 13).

[43] Lyon, D. Making of "Bring Colour to Life" project. Vimeo. https://vimeo.com/29288437 (2011).

Figure 16.6 "Water figures" collection, by Linden Gledhill.
Images courtesy of Linden Gledhill.

and the pain of those back flops at the swimming pool when I was trying to learn to somersault. Each of us may experience a different souvenir. Perhaps Roger Payne would have seen here the colored version of the splashes that humpback whales cause after breaching. Gledhill's pictures can also bring to mind more than a century of inventions powered by microfluidics—asthma nebulizers, fuel injectors, irrigation systems, airbrushes, and inkjet printers— as a testimony of what humans are capable of. They remind us how the world flows.

Summary

- Roger Payne, the man who discovered that whales sing, was also an activist who started a global movement that ultimately saved the whales.
- The inner ear of all vertebrates includes hearing and motion sensors that are highly conserved in evolution and primarily microfluidic in function.
- High-frequency sound, or ultrasound, can be used to separate molecules, particles, and cells in microchannels—a technique called *acoustophoresis* that promises to bypass the need for complex or bulky instruments such as centrifuges still widely used in hospitals and biomedical laboratories.

17

The Microfluidic Race against Cancer

Circulating Tumor Cells, the Microfluidic Circuits of the Lymphatic System, and the Circulating Tumor Cell Chip

Who has not suffered the cruelty of cancer—if not in our own body then in those of loved ones. Leukemia stole my eleven-year-old middle school classmate Fredi away from the playground one day. After a full year of dreadful treatments in a Parisian hospital far from home, our class cried goodbye before his child-sized coffin from the first rows of a gloomy church while the priest promised him eternal life. Almost forty years later, before a much larger casket, I would bid farewell to my dad, looking dignified as ever in his favorite suit, the smile on his face belying the pain his colorectal tumor had made him endure. Cancer wreaks havoc in innumerable ways.[a] Although there is much progress to be done in the fight for cancer, one of the latest advancements comes from a nifty microfluidic device that samples our blood for rogue cancer cells.

Oswald Peterson, an ex-professional Black dancer from Brooklyn, New York, remembers very well the day he was diagnosed with lung cancer. It was in 2017, and he had severe back pain and shortness of breath. Peterson was forty-nine years old then, and after a short visit to the doctor, he had been sent home with antibiotics. As the pain worsened over the day, he took an hour-long Uber ride from his home in Brooklyn to the Columbia University Medical Center in Manhattan, where he was admitted to the hospital. Doctors had to drain fluid from his lungs and heart. Then, they took a biopsy from his lung, a procedure that involves carefully inserting a needle through his skin and punching out a piece of the lung for analysis. The doctors informed him that he had the most common type of lung cancer, called non-small cell lung cancer[1] (NSCLC), which is often not very sensitive to chemotherapy. Unfortunately, the cancer

---

[a] In the United States alone, doctors diagnosed an estimated over 2 million new cancer cases in 2024, and more than 600,000 patients died from it[1]—making cancer the second leading cause of death after heart disease.

[1] American Cancer Society. Cancer facts & figures 2024. https://www.cancer.org/research/cancer-facts-statistics/all-cancer-facts-figures/2024-cancer-facts-figures.html (2024).

had spread to distant sites throughout his body, including his spine, a state that doctors call "stage IV."[2]

For Peterson, the worst part was learning that there are no more stages after stage IV; the estimated five-year survival rate for patients with the highest stage like him, he was told that day, is a dismal 6%. The inexorable spreading of cancer to distant sites that was happening in him is called *metastasis*. Patients rarely die of the first tumor—in his case, in the lung—called the *primary tumor*. More often than not, the primary tumor grows unnoticed and metastasizes to a vital organ before it can be detected. Between 1996 and 2000, researchers conducted an extensive study in the east of England with thousands of cancer patients diagnosed with one of the eight most common cancers. The now-famous study found that 90% of patients diagnosed at stage I survived for at least ten years, compared with only 5% for those diagnosed at stage IV.[3] Early stage treatments are not only more effective but also less costly. In cancer, an early diagnosis is vital, and a late diagnosis is bad news.

Peterson immediately started making arrangements for his funeral—he put together a playlist of songs and even picked out his burial outfit. His mother and his partner had both passed away recently; he was weak and felt depressed with little to live for—but, ironically, poor health played in his favor this time around. The first line of treatment for patients with NSCLC in stage IV is usually chemotherapy, but his oncologist thought his body would not tolerate the side effects of chemo. So she thought he could be a good candidate for a new, still-experimental treatment called *immunotherapy* based on activating the patient's immune cells to fight the tumor.[2] "How about the cost?" he wondered. This treatment was part of a clinical trial, so the pharmaceutical company would foot 100% of the bill. Peterson thought he had nothing to lose in trying.

Immunotherapy is the treatment of a disease such as cancer[b] by activating or suppressing the immune system. The cells of the immune system are incredibly efficient at finding their targets, partly because they have privileged access to a parallel microcirculatory system. Blood is not the only fluid flowing through tiny vessels in our bodies. For example, the liver has a parallel system of microvessels to excrete the bile, and there are other examples. But perhaps the most intriguing

---

[b] Autoimmune diseases and conditions such as allergies can also be treated by immunotherapy.

[2] Konkel, L. A return to Carnival. *Cancer Today*. https://www.cancertodaymag.org/spring2021/a-return-to-carnival (2021, March 24).

[3] Meikle, J. Cancer survival rates three times higher with early diagnosis. *The Guardian* https://www.theguardian.com/society/2015/aug/10/cancer-survival-rates-higher-early-diagnosis (2015, August 10).

of all is the *lymphatic system*.[4] This additional microfluidic circulatory network works in parallel with the blood in all vertebrate animals (Figure 17.1, top). The fluid that circulates in the lymphatic system is the *lymph*. The lymphatic system does not have a heart; instead, the lymph is moved along by pressure from muscle contractions, by pulling into the lung with breathing, and by the *peristaltic* (worm-like) action pumping of tiny adjacent, specialized muscle cells that line lymphatic vessel walls.[5] Special valves in the lymphatic channels called *bicuspid valves* ensure the flow always goes in the same direction (Figure 17.1, bottom).

The lymphatic system serves two primary purposes. First, because the porous walls of blood vessels continually leak plasma into tissues, a substantial amount of *interstitial fluid* between tissues needs to be dealt with—about 3 liters per day in humans. The lymphatic system collects this interstitial fluid around the body and brings it back to the blood—near the heart, to be precise. It's a drainage system. Without a properly functioning lymphatic system, our tissues become bloated with fluid.

Second, white blood cells called *lymphocytes*—the very clever sentinels of our immune system that detect and kill all kinds of cells that can make us sick, including cancer cells—use the lymphatic system as a shortcut to roam our body, passing in and out of the blood as they see fit. Picture lymphocytes as the emergency medical responders of our body: It's as if Nature had reserved ambulance lanes for them to ensure they get to all the flash points of the complex megalopolis of our body as quickly and efficiently as possible. Both the immune system and immunotherapy work efficiently because of the lymphatic system.[c]

After just one cycle of the new medication, Peterson felt a significant improvement. When he was released from the hospital, he felt so good he started exercising and dancing again. Since then, he has had many cycles of immunotherapy. And luckily for him, his body seems to tolerate it very well, which is not the case for many immunotherapy patients.

The downside is that Peterson will need to go to the hospital every six months to have a computed tomography (CT) scan of his lungs to help doctors determine if there has been any new tumor growth. The CT scanner machine (Figure 17.2) takes a three-dimensional X-ray image of the inside of one's body. It's not

[c] To make lymphocytes more efficient, scientists extract them from patients and genetically engineer them to obtain more potent lymphocytes that are then reintroduced into the patient. Microfluidic engineers design microdevices that make these immunotherapy fluid manipulations safer, faster, and with a higher yield.[6]

[4] Lumen Learning. Anatomy of the lymphatic and immune systems. *SUNY–OER Services.* https://courses.lumenlearning.com/suny-ap2/chapter/anatomy-of-the-lymphatic-and-immune-systems (n.d.).

[5] Von Der Weid, P.-Y., & Zawieja, D. C. Lymphatic smooth muscle: The motor unit of lymph drainage. *Int. J. Biochem. Cell Biol.* 36: 1147–1153 (2004).

[6] Kim, H., Kim, S., Lim, H., & Chung, A. J. Expanding CAR-T cell immunotherapy horizons through microfluidics. *Lab Chip* 24: 1088–1120 (2024).

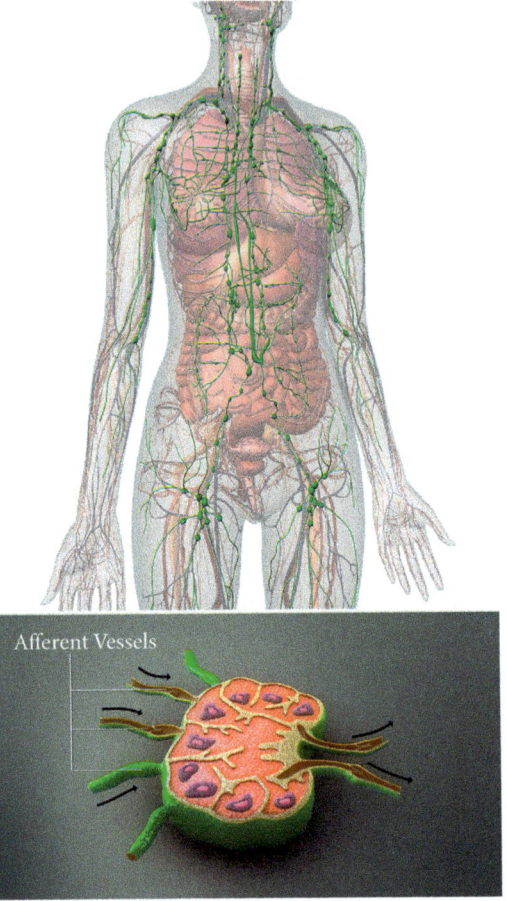

Figure 17.1 (Top) The lymphatic system, shown in green; note the multiple pea-sized lymph nodes. (Bottom) A cross-section of a lymph node, depicting how the bicuspid valves before and after the lymph node allow flow in the direction shown by the arrows; when flow reverses, the valves close and impede flow.

Image sources: (Top) Hank Grebe/Shutterstock. https://www.shutterstock.com/image-illustration/3d-rendering-female-lymphatic-system-1660498840. (Bottom) www.scientificanimations.com. https://en.wikipedia.org/wiki/Lymphatic_vessel#/media/File:3D_Medical_Animation_of_Afferent_Vessel.jpg.

clear how long he will have to take immunotherapy or what will happen if he stops taking it. The uncertainty can be difficult, but "there is only one option and that is to move on from where you are at," he says.[2]

Depending primarily on CT scanners to monitor and manage the care of cancer patients like Peterson is not without its problems. An expert in cancer

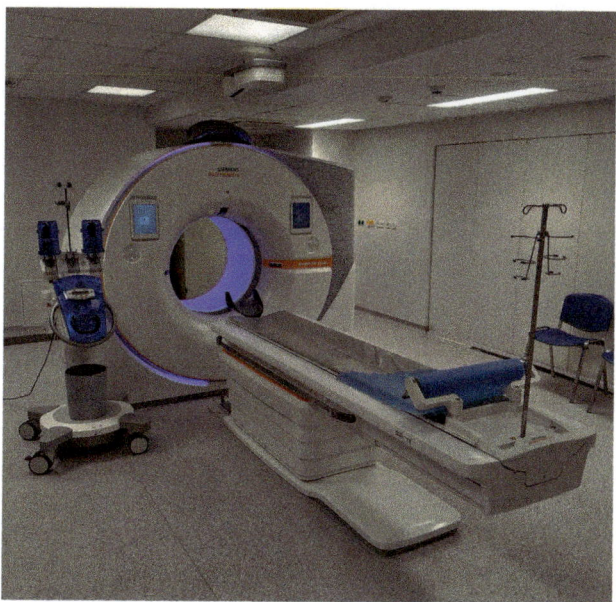

Figure 17.2 CT scanner.

*Image source: User Tomáš Vendiš on Wikipedia. https://en.wikipedia.org/wiki/CT_scan#/media/
File:Modern%C3%AD_v%C3%BDpo%C4%8Detn%C3%AD_tomografie_s_p%C5%99%C3%
ADmo_digit%C3%A1ln%C3%AD_detekc%C3%AD_rentgenov%C3%A9ho_z%C3%A1%C5%
99en%C3%AD.jpg (CC BY-SA 4.0).*

research and treatment such as Dr. Daniel Haber, a Harvard University oncologist who is the Director of the Cancer Center of the Massachusetts General Hospital (MGH), will tell you that CT imaging often misses tumors smaller than 1 millimeter or so in size. It is also expensive, so many insurance companies and hospital administrators limit how often a CT scan can be given—usually only authorizing one every few months. But doctors and patients who suffer from the disease would benefit from a compilation of frequent images that would, in a way, function like a time-lapse movie for their peace of mind. Tumors often become resistant to the treatment, so monitoring with CT scans every six months means a patient loses time and potential new treatment options if cancer starts to grow back—which is bound to happen in most patients.

In May 2004, Haber was introduced to Dr. Mehmet Toner, a Turkish-born bioengineer working across the street at MGH. By his own account, Toner "had not been a good student" in high school. He initially wanted to be a surgeon but could not get into medical school—so he decided to study mechanical engineering in Istanbul instead. In college, Toner became a motivated student. For his

PhD, he got into the Massachusetts Institute of Technology, where he discovered the nascent field of biomedical engineering. "I started reading about it with great interest because of my failure in medicine," says Toner,[7] who obtained a position at Harvard Medical School and MGH. In the early 2000s, he became interested in analyzing "bodily dirty fluids" (blood, plasma, urine, etc.), as he calls them. Unfortunately, at the time, microfluidic devices were too simplistic.

Toner presented Haber with a potential solution to the CT scanner problem that involved a novel microfluidic device. Toner argued that a method based on blood sampling would allow for obtaining a "molecular movie" of the cancer as it evolved during treatment. Blood extraction is simple, inexpensive, and can be repeated often—so doctors could repeat this procedure over time to understand what molecular changes are taking place in the tumor. The detection and analysis of tumor cells in blood could help doctors determine how well the treatment works or if cancer has returned. Perhaps most important, the microfluidic device that Toner envisioned might catch the disease at earlier stages, before the tumor spreads, and improve treatment plans and *prognosis* (a prediction of how the patient's disease will evolve). It could be revolutionary.

Toner explained to Haber that his microfluidic device idea was based on detecting *circulating tumor cells* (CTCs; Figure 17.3), cancer cells shedded by tumors that roam around in the bloodstream in extremely low abundance.[d] Toner had made some quick calculations and summarized all the rare cell detection techniques used until then. This list included technologies such as antibody-coated magnetic beads and shooting cells down through glass capillaries, also called *flow cytometers*. He realized that people were tweaking techniques not intended for scarce cells. "It's like racing against a Formula 1 with a Hyundai—you are not gonna win," he said.[7] In a patient with metastatic disease, every milliliter of blood has approximately one CTC surrounded by a few million white blood cells and a billion red blood cells. Therefore, his device would need to process at least 10 milliliters of blood, about one large blood tube. This volume was considered very large for the available handheld microfluidic devices. To give you an idea, there are an average of 2,000 peas in a pound of peas. Finding one brown pea among half a million 1-pound bags—a truckload—of green peas is equivalent to the challenge that Toner was facing of finding

---

[d] The discovery of CTCs is attributed to Thomas Ashworth, a resident physician at Melbourne Hospital, Australia, who in 1869 looked at the blood of a patient with metastatic cancer under the microscope and observed non-blood cells that were morphologically similar to those in the primary tumor. In his study, he concluded that these "circulating tumor cells" must have traveled from the original tumor site to thirty secondary distant sites.

[7] Folch, A. *Hidden in Plain Sight: The History, Science, and Engineering of Microfluidic Technology.* MIT Press, 2022, 239, 241, 247.

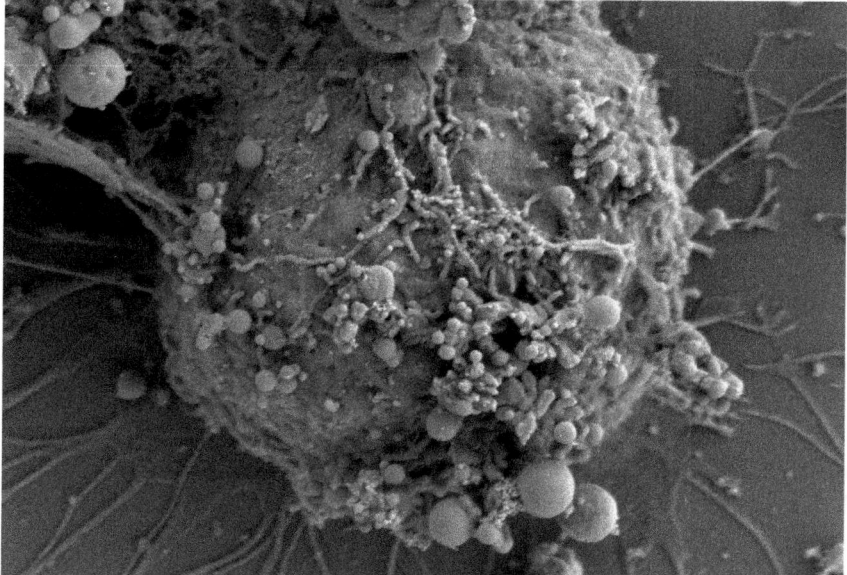

Figure 17.3 Scanning electron micrograph of a circulating tumor cell isolated from the blood of a breast cancer patient. The cell is about 10 micrometers in diameter. *Image courtesy of Shannon Stott, Massachusetts General Hospital.*

CTCs in blood. In short, Toner wanted to search for a needle in a haystack, so he needed the equivalent of "a computer mainframe for microfluidics."[7]

Toner knew from the beginning that the biggest challenge would be getting rid of ("debulking") the other cells—the ubiquitous red blood cells. Our lives depend on these 25 trillion cells[8] that act like a microfluidic mail delivery service for oxygen (see Chapter 15). The only problem for Toner was that the red blood cells outnumbered the rogue cells he was interested in capturing—the CTCs— by *a billion to one.*

Haber was immediately attracted to Toner's groundbreaking approach on two levels. On a personal level, Haber could feel a connection with looking for needles in a haystack. He remembers well when his parents ran a Jewish resettlement organization in Geneva, Switzerland, and the challenges they faced reuniting survivors. At the age of eighteen, Judith Boros had lost all her immediate family when she volunteered to take a trainload of Hungarian orphans to Paris at the end of World War II. In Paris, she met and fell in love with Daniel's father, Irving Haber, a Polish American who, after World War II had returned to

[8] Abbott, A. Scientists bust myth that our bodies have more bacteria than human cells. *Nature* (2016). doi:10.1038/NATURE.2016.19136.

Europe to help resettle Holocaust survivors. They married and had two children before moving their organization to Geneva, where Daniel grew up.

On a professional level, Haber instantly understood that detecting CTCs in the blood would bypass the risk, discomfort, and cost of an invasive needle biopsy and those associated with the subsequent CT scans for follow-up. A lung needle biopsy requires hospitalization, an operating room, and anesthesia—none of which would be necessary to draw blood for a CTC count. Appropriately, Toner's revolutionary test would later be called a *liquid biopsy*.[9]

Toner and Haber quickly sketched a work plan for a device. They would make all the blood cells pass through a microchamber full of microposts. The microposts would be sticky only to CTCs, so the device would "capture" or immobilize the CTCs while letting all the other cells pass through. Microfabricating the chip was not straightforward, so postdoc Sunitha Nagrath optimized the design first with computer modeling. The final design consisted of a triangular array of circular microposts, with each post a tenth of a millimeter in height and width and a gap half of that between microposts. To fabricate a chamber with 78,000 tiny posts, they first used a reactive gas that etched the posts in silicon. Then, they covered the post array with transparent tape to form a chamber. Fortunately for the team, many tumor cells "give away" that they are cancerous by indirectly showing that they originate from the inner linings of tissue (named *epithelia*). They display little ribbons of *epithelial cell adhesion molecule* (EpCAM) on their surface. So Nagrath coated the microposts with an antibody against EpCAM, which made the posts adhesive to CTCs. In 2007, the team ran blood from patients with metastatic cancers through the device—the first of its kind—and demonstrated that it could capture an average of 132 CTCs per milliliter of blood from all tested patients (NSCLC and prostate, pancreas, breast, and colorectal cancers), but not from healthy individuals.[10]

The vast majority of the cells decorated with EpCAM on their surface turned out to be CTCs. However, not all CTCs would be so easy to catch. The team found that some CTCs shredded their EpCAM, and others got covered with platelets, causing the CTCs to miss the antibody-coated posts. So, they decided to build a chip that eliminated all the known normal cell types, and then they analyzed the few remainders.[11] The chip, which fit in a glass slide (Figure 17.4),

[9] Pantel, K., &Alix-Panabières, C. Circulating tumour cells in cancer patients: Challenges and perspectives. *Trends Mol. Med.* 16: 398–406 (2010).

[10] Nagrath, S., Sequist, L. V., Maheswaran, S., Bell, D. W., Irimia, D., Ulkus, L., et al. Isolation of rare circulating tumour cells in cancer patients by microchip technology. *Nature* 450: 1235–1239 (2007).

[11] Mishra, A., Dubash, T. D., Edd, J. F., Jewett, M. K., Garre, S. G., Karabacak, N. M., et al. Ultra-high throughput magnetic sorting of large blood volumes for epitope-agnostic isolation of circulating tumor cells. *Proc. Natl. Acad. Sci. USA* 117: 16839–16847 (2020).

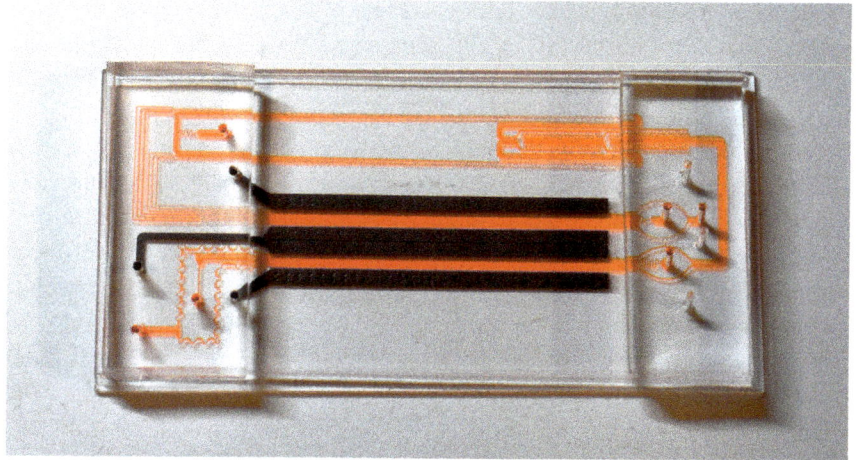

Figure 17.4 Microfluidic platform for ultra-high-throughput isolation of circulating tumor cells from large blood volumes. Despite being only 3.8 centimeters wide and 7.6 centimeters long, this chip can process 1 liter of human blood (containing 6 billion nucleated cells) every hour.
Image courtesy of Avanish Mishra and Mehmet Toner, Harvard Medical School.

worked in stages. The first or "debulking" stage discarded the millions of red blood cells from the few white blood cells and CTCs. In the last step, a magnetic field was applied to separate the white blood cells (magnetically labeled) from the CTCs (which were not labeled). This step allowed for the isolation of CTCs and their genetic analysis.[11] "We pull out all the cells that are not supposed to be in the blood, and then sequence the DNA [of the remaining ones] to find which ones are cancerous," explained Toner.[7]

It is now established that the number of CTCs before (and sometimes during) treatment is a good predictor of the chances of survival of patients with metastatic breast,[12] colorectal,[13] and prostate[14] cancer. In patients with breast cancer, clusters of CTCs (even more rare than single CTCs) appear to have

[12] Cristofanilli, M., Budd, G. T., Ellis, M. J., Stopeck, A., Matera, J., Miller, M. C., et al. Circulating tumor cells, disease progression, and survival in metastatic breast cancer. *N. Engl. J. Med.* 351: 781–791 (2004).

[13] Cohen, S. J., Punt, C. J. A., Iannotti, N., Saidman, B. H., Sabbath, K. D., Gabrail, N. Y., et al. Relationship of circulating tumor cells to tumor response, progression-free survival, and overall survival in patients with metastatic colorectal cancer. *J. Clin. Oncol.* 26: 3213–3221 (2008).

[14] De Bono, J. S., Scher, H. I., Montgomery, R. B., Parker, C., Miller, M. C., Tissing, H., et al. Circulating tumor cells predict survival benefit from treatment in metastatic castration-resistant prostate cancer. *Clin. Cancer Res.* 14: 6302–6309 (2008).

twenty-three- to fifty-fold increased metastatic potential.[15] Compared to other methods that only detect CTCs in 20% of patients with advanced metastatic disease, Toner's CTC chips reach detection in 80–100% of patients, depending on the cancers. The diagnostic test only requires detecting a few (three to five) CTCs from 20 milliliters of blood (a human has an average of 5 liters). Toner hopes that his CTC chips will soon enable the collection of highly viable CTCs from ever-larger blood volumes in the clinic, and perhaps make early cancer detection and monitoring the new walk-in clinic blood test for everyone. These ultra-high-throughput microfluidic chips are the Formula 1 equivalent of microfluidics, the fast-paced technology for a near future in which I would have loved to see Fredi and my dad.

Summary

- All vertebrate animals have a microfluidic circulatory network that works in parallel with the blood called the *lymphatic system.*
- Lymphocytes are clever sentinels of our immune system that detect and kill cancer cells and use the lymphatic system as a shortcut to roam our body, going in and out of the blood as they see fit.
- A form of cancer therapy called immunotherapy uses the patient's own immune system to fight the tumor cells.
- Circulating tumor cells shed from primary tumors and are found in the blood of cancer patients at a concentration of a few CTCs per billion red blood cells.
- A microfluidic chip pioneered by Mehmet Toner and Daniel Haber could isolate large numbers of CTCs from almost all tested patients (NSCLC and prostate, pancreas, breast, and colorectal cancers). This approach promises the painless and sensitive monitoring of cancer progression and, in the near future, possibly the sensitive detection of cancer spread before it can even be detected by imaging.

---

[15] Aceto, N., Bardia, A., Miyamoto, D. T., Donaldson, M. C., Wittner, B. S., Spencer, J. A., et al. Circulating tumor cell clusters are oligoclonal precursors of breast cancer metastasis. *Cell* 158: 1110–1122 (2014).

18

The Innocence of DNA

The Microfluidic DNA Sequencing Chips and Their Impact on Society

Marvin Anderson is a soft-spoken, educated Black man who manages a gentle smile behind his frameless glasses as he tells the story of how maliciously a police officer, a lawyer, and a judge—all White men—conspired to rob him of the best years of his life.[1] It took him almost two decades—and the emergence of modern DNA testing, a set of techniques supported by microfluidic technology—to prove his innocence.

Growing up in Virginia, like kids everywhere, Anderson dreamed of becoming a firefighter. As a young man, he was able to enter the academy, where he started studying to become a professional firefighter. Then, on July 17, 1982, a young woman was raped in his neighborhood. She reported the rapist as being a total stranger who was riding a bicycle—which turned out to be stolen—and who had told her that he "had a White girl."

This is where Anderson's nightmare started. A manipulative and obsessive police officer singled out Anderson as a suspect because Anderson was the only Black man he knew who was living with a White woman. The police officer then prepared a photo lineup followed by an in-person lineup designed to incriminate Anderson. The photo lineup consisted of a color photo of Anderson and six black-and-white images of Black men, so the victim identified Anderson. In the in-person lineup, Anderson was the only person who was also in the photo lineup, so the victim "recognized" him again.

Luckily for Anderson, the bicycle owner quickly identified the bicycle and the thief—Mr. John Otis Lincoln. But, inexplicably, Anderson's lawyer ignored his plea to call Lincoln (or the bicycle owner) to testify, which dug him deeper into his Kafkaesque ordeal. An all-White jury found Anderson guilty of rape, abduction, sodomy, and robbery. On December, 14, 1982, at the age of eighteen, Anderson was sentenced to 210 years in prison.

About five and a half years later, in August 1988, John Otis Lincoln came forward and, in a hearing in open court under oath, confessed the details of

---

[1] Innocence Project. Marvin Anderson. https://innocenceproject.org/cases/marvin-anderson (2024).

the crime. It should have been the end of Anderson's nightmares. Indignantly, the same judge who had convicted Anderson refused to vacate the conviction.[1]

In the late 1980s, when Anderson was incarcerated, there was no DNA technology to identify individuals biologically. Because each human has unique DNA, DNA genotyping (the unique pattern of DNA sequences at particular locations) and sequencing have provided an essential tool that has dramatically improved the reliability of forensic science in the past thirty years. DNA can be isolated and amplified from a fingerprint, a hair, a speck of saliva, or a spot in a handkerchief. Our DNA is the molecular blueprint that contains the instructions with which our bodies are built in the womb and then repaired and operated daily. This blueprint is a unique identification tag, although it is not as easy to read as this sentence. It contains 3 billion "letters" of DNA—A, C, G, or T—the initials of chemical groups or DNA building blocks called *nucleotide bases*: adenine, cytosine, guanine, and thymine. The DNA molecule has the shape of a ladder twisted as a double helix (Figure 18.1). Inside the double helix, two nucleotide bases form each of the rungs of the DNA ladder and encode the genetic information. The information on one helix strand always mirrors the other strand, with the bases always matched as the same pairs: A always mirrors T, and C always mirrors G, so only one strand needs to be read. DNA sequencing technology has allowed scientists to decode the meaning of those instructions, from the paths to disease to those of evolutionary change, differentiating mouse from human and revealing the Neanderthal DNA living on within us. This achievement has revolutionized our lives and occurred within the lifetime of most readers of this book. These tests would not have been possible without the development of microfluidic sequencers in the 1990s.

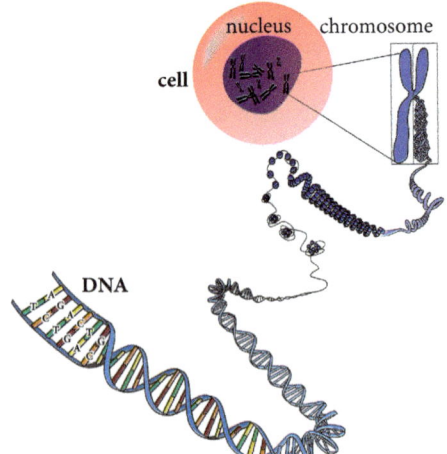

Figure 18.1 The DNA molecule.

Image source: User OpenClipart-Vectors on Pixabay. https://pixabay.com/vectors/genetics-chromosomes-rna-dna-156404/.

One of the key developments was the miniaturization of the first stage of DNA sequencing by microfluidics. The analysis of tiny samples of DNA starts with its "amplification" by artificial replication until it is available in measurable quantities. This replication reaction is called polymerase chain reaction (PCR). PCR is a method to exponentially make many copies of (or "amplify") a given DNA molecule. This technique has been so transformative for modern biology that its inventors, Kary Mullis and Michael Smith, were awarded the Nobel Prize in 1993. The method first uses heat to separate the DNA double helix into its two strands. Starting with a template that binds to one of the strands (the "*primer*"), a natural enzyme called *DNA polymerase* then rebuilds the double helix by adding one nucleotide base at a time, effectively building a mirror image of the original strand much like a zipper closes. Thus, each heat cycle results in the duplication of the number of DNA molecules. After, for example, ten cycles, the number of molecules has multiplied by $2^{10} = 1,024$, resulting in rapid, exponential amplification of DNA material. Because chemical reactions are more controlled and faster in small volumes, engineers developed microfluidic PCR chips to expedite DNA processing (Figure 18.2).[2,3]

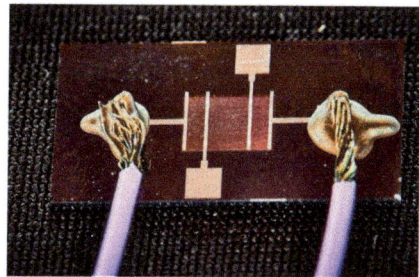

Figure 18.2 (Left) The first microfluidic PCR chip, fabricated at the Lawrence Livermore National Lab in California by Allen Northrup circa 1993. Only the base of the device is shown (the microchannels, which go on top, are not shown). The electrodes serve to apply the heat cycles. (Right) Continuous-flow PCR chip, made at the Imperial College in London by Andreas Manz's group circa 1997. The DNA sample flows through regions of hot and cold substrate, Hence, the timing of the heat cycles is determined by the flow itself.
Images courtesy of CBMS (Okaar Photography).

[2] Northrup, M. A., Ching, M. T., White, R. M., & Watson, R. T. DNA amplification in a micro-fabricated reaction chamber. In *Transducers '93: The 7th International Conference on Solid-State Sensors and Actuators, June 7–10, 1993, Yokohama, Japan*. Institute of Electrical Engineers of Japan, 1993.

[3] Kopp, M. U., Mello, A. J., & Manz, A. Chemical amplification: Continuous-flow PCR on a chip. *Science* 280: 1046–1048 (1998).

DNA amplification by PCR became the critical step for in-depth analysis of DNA. One such application is *genotyping*, or the characterization of which genetic type an individual has among the different possibilities, analogous to one's blood type. Genotyping can be done by indirect measures of DNA sequence, such as using "satellite markers" (done by PCR), or by direct *sequencing* (the "reading" of the "letters" of DNA). Instead of reading letter by letter, satellite markers only match the lengths of stereotypical sequence fragments as a way to characterize the DNA. In addition to genotyping, PCR amplification of DNA made it possible to read the sequence of DNA easily and efficiently in subsequent procedures, a transformative development.

The sequence of DNA letters forms our *genes*, the genetic information that makes us different. The complete set of our genes is 3 billion letters long and is called our *genome*. The first human genome ever read[a] required a $3 billion, 13-year-long international effort from 1990 to 2003 called the Human Genome Project. It may seem a bit pricey now for just one genome, but it spurred a scientific revolution.

Every sequencing technique used for and developed since the Human Genome Project has hinged on microfluidics. And not just for the PCR chips but for the readout as well. Reading a human genome for the first time was a long, tedious process based on amplifying the DNA, chopping the genes into multiple fragments, sorting them by size, and reassembling them by overlapping sequences. It was like trying to reassemble a book after running multiple copies of the book through a shredder, picking one piece of paper at a time. The size sorting of DNA fragments was done with an old microfluidic technique called[4] *capillary electrophoresis*.[5,b] The DNA fragments were introduced into tiny glass capillaries, and a voltage was applied between the ends to separate the fragments by charge. Because the amount of charge in a DNA molecule is proportional to its size, the pieces appeared at the other end of the microchannel in order of their mass. Since DNA by itself is invisible, the presence of each "letter" (A, C, G, or T) had to be visualized by different fluorescent molecular labels. This was a precise process, but it wasn't fast.

The parallelization and integration of capillary electrophoresis into microfluidic chips contributed to speeding DNA genotyping and DNA

---

[a] The volunteers who donated their DNA were de-identified so we will never know whose DNA was the first to be sequenced. In contrast, in 2002 the source of the DNA sequenced by Celera, the private competitor of the publicly funded Human Genome Project, was revealed to be its founder, Craig Venter. In the end, both shared the credit for the successful sequencing project.[4]

[b] Capillary electrophoresis was invented by James W. Jorgenson and Krynn DeArman Lukacs in 1981,[5] before the term microfluidics was coined.

[4] Wade, N. Scientist reveals secret of genome: It's his. *The New York Times.* https://www.nytimes.com/2002/04/27/us/scientist-reveals-secret-of-genome-it-s-his.html (2002, April 27).

[5] Jorgenson, J. W., & Lukacs, K. D. A. Zone electrophoresis in open-tubular glass capillaries. *Anal. Chem.* 53: 1298–1302 (1981).

sequencing. In 1992, when the Human Genome Project desperately sought solutions to increase the throughput of its slowly advancing DNA sequencing efforts, a PhD student named Adam Woolley joined the lab of Richard Mathies, a physical chemist at the University of California, Berkeley. They immediately hit it off. Woolley, a smart student from a family of chemists, had been accepted by the university on a fellowship. Mathies presented him with a clear proposal of how miniaturization would help in DNA analysis. Woolley built a capillary electrophoresis chip that could separate DNA fragments at ultra-high speeds in glass microfluidic channels filled with a sieving matrix made of cellulose derivative.[6] To increase processing speed, Mathies and Woolley integrated PCR chips with the channels;[7] the group later scaled up the process to 384 channels in parallel (Figure 18.3).[8] This research ultimately led to the commercialization in 1998 of the MegaBACE DNA sequencer, one of the two primary DNA sequencing machines used by groups throughout the world to complete the human genome.

In 1997, amid the Human Genome Project efforts, Shankar Balasubramanian and David Klenerman, chemists at the University of Cambridge (UK), decided

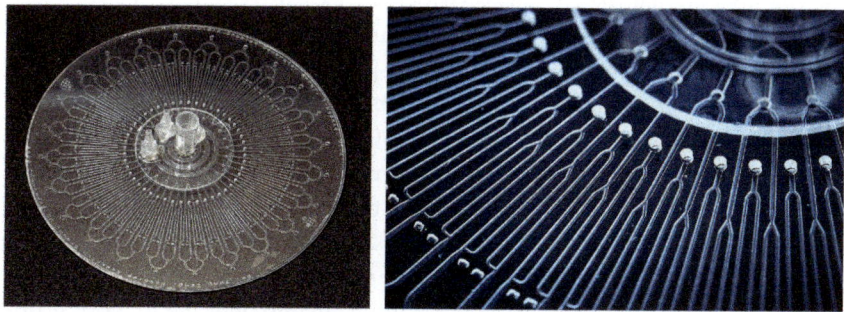

Figure 18.3 (Left) A 384-channel DNA sequencer. The disk-shaped device is 10 centimeters in diameter and etched in glass. At the inner and outer circles, 384 pairs of electrodes are introduced to apply the voltage needed for on-chip capillary electrophoresis. Fluorescence readout is performed on the outer circle using a custom-designed microscope. (Right) Detail of the above.
Images courtesy of CBMS (Okaar Photography). Devices by Richard Mathies' group.

[6] Woolley, A. T., Sensabaugh, G. F., & Mathies, R. A. High-speed DNA genotyping using microfabricated capillary array electrophoresis chips. *Anal. Chem.* 69: 2181–2186 (1997).
[7] Woolley, A. T., Hadley, D., Landre, P., DeMello, A. J., Mathies, R. A., & Northrup, M. A. Functional integration of PCR amplification and capillary electrophoresis in a microfabricated DNA analysis device. *Anal. Chem.* 68: 4081–4086 (1996).
[8] Emrich, C. A., Tian, H., Medintz, I. L., & Mathies, R. A. Microfabricated 384-lane capillary array electrophoresis bioanalyzer for ultrahigh-throughput genetic analysis. *Anal. Chem.* 74: 5076–5083 (2002).

they did not want to wait for a decade or so for every human genome. They realized the key would be to ditch the shredder-and-pick approach. They were dreaming of a sequencing method so fast and inexpensive it could save the lives of patients with cancer and many other diseases, assist people in Marvin Anderson's situation, or simply help folks with paternity claims and genealogy pastimes. The clue of how to do just that occurred to them on August 26, 1997, while sitting in the beer garden of Cambridge's pub, Panton Arms.

Their revolutionary technique was called *next-generation sequencing* (NGS). Balasubramanian and Klenerman continued to use shredding, but instead of carefully picking one piece at a time, they devised a way to read all the pieces simultaneously. They shredded more than one copy of a genome, so assembling it was a matter of finding the overlaps between the fragments: Piecing together a book from several shredded copies is easier than if only one copy had been shredded.[c]

Their NGS idea was based on "sequencing by synthesis," which consisted of placing all the shreds of DNA on a surface and chemically reacting them with a series of fluorescent compounds of different colors that could unveil the DNA sequence by taking pictures. All the reactions occurred in a microfluidic chamber so that the chemicals (which are expensive) could be exchanged automatically and very quickly while consuming a small volume of reagents. The floor of the microfluidic chamber was printed with primers—short DNA molecules that jump-started the DNA binding reaction—to which the shredded pieces attached. Then, each shred (about 100 letters long) was amplified on the surface until it formed a cluster of (for a modern microscope) approximately 1,000 identical copies. The clusters are like the mold on your bread—you cannot see them until they grow to a specific size.[d] Once the clusters were ready (Figure 18.4, left), the sequencing of each shred began.

The rest was simply a matter of putting the chemistry of life to work inside a microfluidic chamber so it could be automated and done with a relatively small volume of reagents. The letters of DNA were invisible, so the researchers washed the surface with a cocktail of the four nucleotide bases A, T, C, and G, each fluorescently tagged in a different color and attached to their phosphate

---

[c] To assemble a whole genome becomes complicated because there is a great abundance of repetitive sequences in different parts of the genome, so in practice, this strategy requires repeating the process several times to ensure the fragments overlap.

[d] In most chips, the clusters are distributed at random and form so-called polyclonal clusters with more than one sequence. In the latest-generation chips, the clusters are arrayed in nanowells, forming "monoclonal" clusters that originate from a single sequence, which makes the whole process more precise (but also more expensive).[9]

[9] Illumina. Illumina: Patterned flow cell technology. YouTube. https://www.youtube.com/watch?v=pfZp5Vgsbw0 (2015).

[10] Ibiology Techniques. Next generation sequencing 1: Overview—Eric Chow (UCSF). YouTube. https://www.youtube.com/watch?v=mI0Fo9kaWqo (2019).

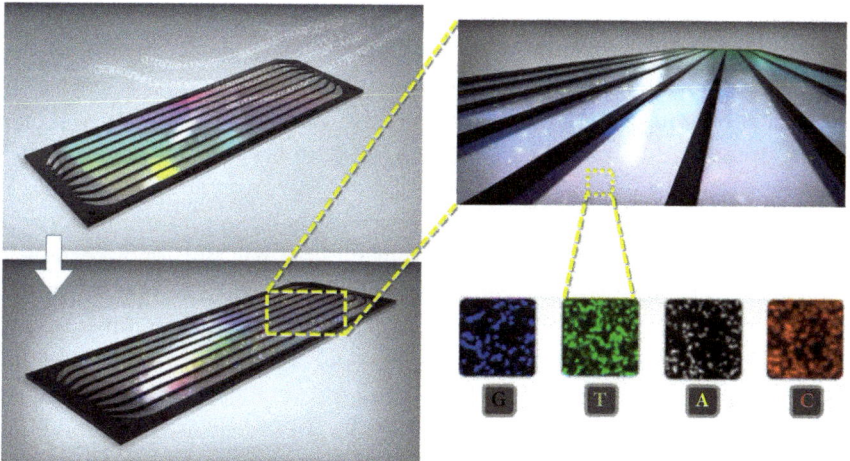

Figure 18.4 DNA sequencing process in a microfluidic chamber.

Image sources: All images obtained from screenshots of YouTube videos.[9,10] *https://www.youtube. com/watch?v=pfZp5Vgsbw0 and https://www.youtube.com/watch?v=mI0Fo9kaWqo. Fair use.*

and sugar groups to build the DNA backbone. Here is where DNA synthesis happened, nucleotide by nucleotide. For example, T was labeled green, A was yellow, C was red, and G was blue. Hence, clusters that ended in A lit up in green when the green-labeled T nucleotide bases recognized them, and so on. The researchers could decode an additional letter of the shredded pieces by taking color images of all the clusters for every cocktail wash in the microchamber (Figure 18.4, right). The shreds were about 100 letters long, so after approximately 100 repetitions, no more letters were left to decode.[10] The process took a few days, but not years, and it did not cost billions of dollars because a robot did it with a small amount of fluids. A computer assembled the book-genome once all the shredded pieces had been read. Microfluidics had triumphed.

In an even newer "nanofluidic" technology called *nanopore sequencing*, DNA molecules pass through a molecular-sized pore—like a thread through the eye of a needle. This approach directly reads the DNA sequence without amplification and without labeling. As the DNA passes through the pore, the device measures the amount of electrical current that passes through the pore and registers an electrical signature that is a function of which DNA "letter" is crossing the pore at that instant. This method offers the advantage of real-time analysis and portability. During the COVID-19 pandemic, a quarter of all the viral COVID-19 genomes were sequenced with nanopore devices. With the newer machines, sequencing in remote areas with limited resources or even by high school students is possible.

DNA technology finally ended Marvin Anderson's misfortune. Anderson belongs to the first generation of unjustly incarcerated people who have benefitted from DNA technology, albeit late. The Innocence Project, an organization that works to overturn wrongful convictions, accepted his case in 1994. Although released on parole in 1997, he wanted to keep fighting to prove his innocence. He knew it was not going to be easy because the rape kit and its contents had been destroyed long ago. Fortunately for Anderson, the criminalist who had performed the blood exam in 1982 had not followed policy and had intentionally failed to return some used swabs to the rape kit in the hopes that DNA testing would be possible in the future. Sperm samples recovered from the victim's body, among other physical evidence, were located in the criminalist's laboratory notebook. In 2001, after several denied requests, DNA identification tests were performed. At the time, police only used a genotyping system that searched for satellite markers in thirteen to twenty genetic locations, of which only four were readable in Anderson's degraded sample. This approach can rule out a suspect when there is a mismatch and gives probabilities of whether it is the same person when it is a match. By comparison, the newer DNA genotyping technologies developed with microfluidics examine more than half a million locations for specific single-base changes. These services, such as Ancestry.com or 23andMe (known for health and ancestry data, and 39 million people sequenced together), help police match DNA for crimes absolutely and even to discover suspects with the help of genealogies, famously identifying the notorious Golden State killer in 2018.[e]

The results did not match Anderson's DNA. But they matched two inmates in the Virginia prison system, one of whom was the man who had already confessed to the crime. The cruelty of scientific progress is that this speck of information stayed hidden for nearly two decades before technology was able to unveil it. On the positive side, it finally turned the relentless wheels of Justice in Anderson's favor. On August 21, 2002, Anderson was granted a full pardon by the governor of Virginia. He had spent fifteen years behind bars and four years on parole. After his release, he fulfilled his dream to become a firefighter, and he is now a fire department chief and serves on the board of directors for the Innocence Project.[1]

---

[e] Forensic DNA technology can also be applied to fishing and logging management. Researchers at the University of Washington in Seattle developed an inexpensive and portable DNA analysis device to help a police patrol in the field quickly identify illegally sourced timber before it is turned into furniture or illicitly caught fish before it is canned or turned into oil.[11]

[11] Holmes, H. R., Gomez, A. E., Baisch, D. A., & Böhringer, K. F. Automated species identification device for conservation biology. In *International Conference on Miniaturized Systems for Chemistry and Life Sciences (MicroTAS)*. Chemical and Biological Microsystems Society, 2017.

Between 1992 and 2022, the Innocence Project used DNA to exonerate 375 people who wrongfully served an average of fourteen years in prison each in the United States—a total of 5,250 years of incarceration of innocent people, twenty-one of whom were death row inmates. Crucially, DNA genotyping is increasingly accepted as a legal tool worldwide: The judicial systems of more than 100 countries now use forensic DNA analysis.[12] Modern DNA testing has touched not only these exculpated people and the lives of those whose DNA was directly tested but also all our lives—by changing the legal meanings of culpability, innocence, and impunity.

* * *

The Human Genome Project awakened a ferocious appetite in biomedical researchers to analyze more genomes, cells, and organisms, but most biologists still used manual pipettes to move fluids around in their day-to-day experiments. All this sequencing and analysis frenzy demanded automation and parallelization. In addition to droplet microfluidics (see Chapter 5), which underlie some of the newest innovations in gene sequencing, *microfluidic valves* that control flow in microchannels empowered applications from genetics to protein and cell analysis. To free biologists from their "tyranny of pipetting,"[13] in 2000 Stephen Quake and Marc Unger at Caltech invented a microvalve that could be easily fabricated and operated in arrays with the rest of a microfluidic device.[14]

Quake and Unger built the microvalve as a "sandwich" of a thin, flexible membrane between two channels crisscrossing one on top of the other. The bottom channel carried the fluid flow, and the top ("control") channel contained air to control the valve. Air pressure applied to the control channel deflected the membrane down to occlude the fluid channel below, close the valve, and stop the flow (Figure 18.5, top). Groups of valves were used to construct micropumps[15] and multiplexers.[16]

Because most experimental protocols in biology relied on tedious pipetting and costly reagents, automation had obvious benefits. Quake's group exploited

[12] Wallace, H. M., Jackson, A. R., Gruber, J., & Thibedeau, A. D. Forensic DNA databases—Ethical and legal standards: A global review. *Egypt. J. Forensic Sci.* 4: 57–63 (2014).

[13] Quake, S. Solving the tyranny of pipetting. https://arxiv.org/abs/1802.05601 (2018).

[14] Unger, M. A., Chou, H. P., Thorsen, T., Scherer, A., & Quake, S. R. Monolithic microfabricated valves and pumps by multilayer soft lithography. *Science* 288: 113–116 (2000).

[15] Chou, H.-P., Unger, M. A., & Quake, S. R. A microfabricated rotary pump. *Biomed. Microdevices* 3: 323–330 (2001).

[16] Thorsen, T., Maerkl, S. J., & Quake, S. R. Microfluidic large-scale integration. *Science* 298: 580–584 (2002).

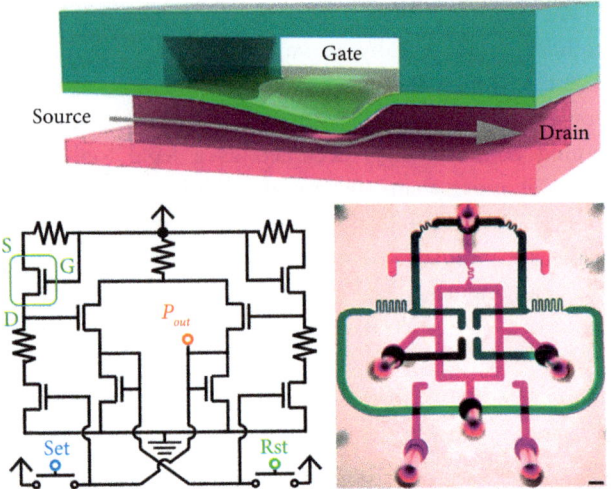

Figure 18.5 (Top) Microfluidic valve. The top channel acts as a "control channel" or "gate" to deflect the membrane (green) against the flow between source and drain in the fluid channel. (Bottom) Schematic (left) and micrograph (right) of an SR-latch, a memory storage device.

Images courtesy of Kaustav Gopinathan and Mehmet Toner.

the microfluidic valves to automate and parallelize, in miniature format, every conceivable molecular biology protocol, ranging from nucleic acid isolation,[17] amplification,[18,19] and analysis[20] to cell sorting[21] and protein crystallography,[22] among others. Other teams applied microvalves to quickly profile stem cells,[23] sort worms,[24] run large-scale immunoassays for cancer diagnostics,[25] and test

[17] Hong, J. W., Studer, V., Hang, G., Anderson, W. F., & Quake, S. R. A nanoliter-scale nucleic acid processor with parallel architecture. *Nat. Biotechnol.* 22: 435–439 (2004).

[18] Marcus, J. S., Anderson, W. F., & Quake, S. R. Parallel picoliter RT-PCR assays using microfluidics. *Anal. Chem.* 78: 956–958 (2006).

[19] Ottesen, E. A., Hong, J. W., Quake, S. R., & Leadbetter, J. R. Microfluidic digital PCR enables multigene analysis of individual environmental bacteria. *Science* 314: 1464–1467 (2006).

[20] Marcus, J. S., Anderson, W. F., & Quake, S. R. Microfluidic single-cell mRNA isolation and analysis. *Anal. Chem.* 78: 3084–3089 (2006).

[21] Fu, A. Y., Spence, C., Scherer, A., Arnold, F. H., & Quake, S. R. A microfabricated fluorescence-activated cell sorter. *Nat. Biotechnol.* 17: 1109–1111 (1999).

[22] Hansen, C. L., Skordalakes, E., Berger, J. M., & Quake, S. R. A robust and scalable microfluidic metering method that allows protein crystal growth by free interface diffusion. *Proc. Natl. Acad. Sci. USA* 99: 16531–16536 (2002).

[23] Zhong, J. F., Chen, Y., Marcus, J. S., Scherer, A., Quake, S. R., Taylor, C. R., & Weiner, L. P. A microfluidic processor for gene expression profiling of single human embryonic stem cells. *Lab Chip* 8: 68–74 (2008).

[24] Chung, K., Crane, M. M., & Lu, H. Automated on-chip rapid microscopy, phenotyping and sorting of *C. elegans*. *Nat. Methods* 5: 637–643 (2008).

[25] Garcia-Cordero, J. L., & Maerkl, S. J. A 1024-sample serum analyzer chip for cancer diagnostics. *Lab Chip* 14: 2642–2650 (2014).

tissue biopsies[26] in numbers that were unthinkable until then. In about a decade, Quake's microfluidic valves converted molecular and cell biology into something that was routinely done by microfluidic machines and in reagent quantities as small as a few billionths of a liter (a *nanoliter*).

A key advantage of a microfluidic valve is that it can work like a (fluidic) transistor. Many microfluidic engineers have used the functional similarity between the microvalve (a fluidic switch) and the transistor (an electronic switch) to build "microfluidic logic gates."[27-33] Kaustav Gopinathan and Mehmet Toner at Massachusetts General Hospital[34] have recently figured out how to operate a valve in such a way that it would switch *and* amplify—that is, they have built a *microfluidic transistor* (see Figure 18.5, top).[f] With the microfluidic transistor, they have been able to build more complex logic circuits, such as an SR-latch (a digital memory storage device; see Figure 18.5, bottom); analog circuits, such as a pressure amplifier; and circuits that combine both digital and analog elements, such as a "smart" single-cell dispenser.[35] With this microfluidic transistor, the possibilities appear endless.

---

[f] Note that any automation technology relies on switches, but to be precise, it relies on signal processing.[38] Historically, this signal processing was done by analog systems, such as water clocks and centrifugal (steam engine) governors (that regulate the amount of steam admitted into the cylinder), which could perform arithmetic operations on analog quantities like torque, flow, and voltage. After the advent of microelectronics, it was found that instead of performing mathematical operations such as addition and multiplication on analog signals, one could approximate similar behavior using a great number of binary logic gates executing on digital signals. This principle is one of the key principles of digital computers today, which use billions of transistors as electronic switches to build binary logic gates. But there is so much more to signal processing than digital logic, and there is much more to a transistor (which can amplify in states that are neither fully open nor fully closed) than a simple switch.

[26] de Hoyos-vega, J. M., Gonzalez-Suarez, A. M., & Garcia-Cordero, J. L. A versatile microfluidic device for multiple ex vivo/in vitro tissue assays unrestrained from tissue topography. *Microsystems Nanoeng.* 6: 40 (2020).

[27] Takao, H., Ishida, M., & Sawada, K. A pneumatically actuated full in-channel microvalve with MOSFET-like function in fluid channel networks. *J. Microelectromechanical Syst.* 11: 421–426 (2002).

[28] Grover, W. H., Ivester, R. H. C., Jensen, E. C., & Mathies, R. A. Development and multiplexed control of latching pneumatic valves using microfluidic logical structures. *Lab Chip* 6: 623–631 (2006).

[29] Wehner, M., Truby, R. L., Fitzgerald, D. J., Mosadegh, B., Whitesides, G. M., Lewis, J. A., & Wood, R. J. An integrated design and fabrication strategy for entirely soft, autonomous robots. *Nature* 536: 451–455 (2016).

[30] Jensen, E. C., Grover, W. H., & Mathies, R. A. Micropneumatic digital logic structures for integrated microdevice computation and control. *J. Microelectromechanical Syst.* 16: 1378–1385 (2007).

[31] Rhee, M., & Burns, M. A. Microfluidic pneumatic logic circuits and digital pneumatic microprocessors for integrated microfluidic systems. *Lab Chip* 9: 3131–3143 (2009).

[32] Mosadegh, B., Kuo, C.-H., Tung, Y.-C., Torisawa, Y., Bersano-Begey, T., Tavana, H., & Takayama, S. Integrated elastomeric components for autonomous regulation of sequential and oscillatory flow switching in microfluidic devices. *Nat. Phys.* 6: 433–437 (2010).

[33] Duncan, P. N., Ahrar, S., & Hui, E. E. Scaling of pneumatic digital logic circuits. *Lab Chip* 15: 1360–1365 (2015).

[34] Bennett, S. A brief history of automatic control. *IEEE Control Syst.* 16: 17–25 (1996).

[35] Gopinathan, K. A., Mishra, A., Mutlu, B. R., Edd, J. F., & Toner, M. A microfluidic transistor for automatic control of liquids. *Nature* 622: 735–741 (2023).

Powered by microfluidic technologies such as capillary electrophoresis, the microvalves, and the droplets, the many ramifications of the Human Genome Project are still helping researchers identify mutations linked to major diseases such as cancer, heart disease, and diabetes, as well as mutations linked to rare illnesses; assess risks of genetic syndromes in the fetus; and develop more efficacious treatments. A fascinating new chapter in the book of Biology is being written by microfluidically driven DNA technology. Genes, proteins, cells, and even tiny organisms are analyzed at incomparable speeds and in colossal numbers that challenge our comprehension. Using the latest generation of microfluidic DNA sequencers, when researchers detect a new variant of a virus or a new mutation of a bacterium, they can obtain its sequence the next day from a saliva swab or a blood prick from the patient—and the new sequence can give doctors a lot of clues on the severity of the infection and how to treat it. In 2015, the fastest genetic diagnosis listed in the *Guinness Book of World Records* was performed by a team led by Dr. Stephen Kingsmore at Rady Children's Institute for Genomic Medicine in San Diego, California; the team read 3 billion bases of a newborn's genome, plus the parents' genomes, then compared them and identified their rare genetic disorder in just twenty-six hours.[36] Scientists have sequenced the genomes of most species, including those of dangerous germs such as the tuberculosis bacterium or the COVID-19 virus, as well as those of extinct species, such as dinosaurs and other fellow *Homo sapiens* such as Neanderthals. We now know that, on average, modern humans in Europe and Asia carry nearly 2% of Neanderthal DNA, and people in East Asia carry as high as 4%.[37] Microfluidics has helped us live longer and healthier lives and contributed to explaining something as profound as our identity.

In the past twenty years, spurred in part by microfluidic miniaturization and automation, sequencing costs have dropped about one-million-fold, stimulating many creative uses of the technology. People such as Marvin Anderson can now fight for their innocence with more precise genetic tools. This is just the beginning. A handheld nanopore sequencer allowed my daughter's public high school genome science class not only to get thrilled about the technology but also to conduct a real study about the sudden surge of an unexpected bacteria in the local Puget Sound water. A sample of stool or water can reveal a

---

[36] Miller, N. A., Farrow, E. G., Gibson, M., Willig, L. K., Twist, G., Yoo, B., et al. A 26-hour system of highly sensitive whole genome sequencing for emergency management of genetic diseases. *Genome Med.* 7: 100 (2015).

[37] Quilodrán, C. S., Rio, J., Tsoupas, A., & Currat, M. Past human expansions shaped the spatial pattern of Neanderthal ancestry. *Sci. Adv.* 9: eadg9817 (2023).

microbiome teeming with the millions of germs living in our gut or the ocean. Microfluidics-driven technology has told us much—and keeps telling us—about who we are and who we were, and it is helping us better understand how life, all its species, our civilization, and our cultures have evolved together on Earth.

This is how the world flows. If we look close, we will see that the soil, the clouds, the rainbows, the plants, even *we* contain microscopic fluids in continuous motion. If Nature spread the seeds of life in tiny droplets across continents and engineered the ascent of fluids in plants using only the energy stored on the surface of water; if She evolved various separate networks of capillaries to supply the cells of large animals with oxygen, nutrients, and immune fighters as needed, and then endowed those organisms with the ability to hear sound and feel movement through microfluidic sensors; if Nature gifted us with sweat as a microfluidic coolant that made us kings of the savannah; and if She made us microfluidic all inside and surrounded us with microfluidic phenomena, why should it surprise us that we, too, came up with microfluidic inventions to dry our wounds, to read in the dark, to paint and perpetuate our knowledge, to power our engines and heal our lungs, to test our bodily fluids for signs of disease, and, finally, to crack open the candid secret of the microscopic life all around us—and in us?

Summary

- The Human Genome Project was carried through largely with DNA sequencers based on *capillary electrophoresis*, a microfluidic technique for separating chemical compounds in a glass capillary using voltages.
- *Next-generation sequencing* (NGS), invented in 1997, is a massively parallel DNA sequencing method based on DNA synthesis in a microfluidic chamber.
- NGS has made DNA sequencing faster and cheaper, thus revolutionizing all fields touched by genomics: cell and molecular biology, evolutionary biology, forensics, and anthropology, among other areas.
- The microfluidic valve, invented by Stephen Quake and Marc Unger in 2000, has allowed for the miniaturization and automation of virtually all molecular and cell biology protocols.

Glossary

acequia: A canal that helps distribute snow runoff or river water to distant fields for irrigation, typically made in dirt by simple digging.

acoustophoresis: A technique that uses ultrasound to separate molecules, particles, and/or cells in fluids, typically in microchannels.

adenine: *See* **nucleotide base.**

aerosol: A suspension of fine solid particles or liquid droplets in air or another gas.

aerosol can: *See* **aerosol spray can.**

aerosol spray can: A device that generates a spray of droplets through a nozzle using pressurized gas. The droplets form spontaneously once the liquid has crossed the nozzle at high speed under the Rayleigh instability.

aichmophobia: Phobia of needles.

airbrush: An air-operated tool that sprays paint, ink, or dye. The first airbrush design, named an "atomizer" by Francis E. Stanley in 1876, is considered the precursor of today's airbrushes based on the Venturi effect. Other airbrushes directly pressurize the fluid through a nozzle and rely on the Rayleigh instability to break the fluid jet into a spray. The invention of the first of these airbrushes is attributed to Joe Binks of Chicago in 1887.

airways: A part of the respiratory system through which air flows.

alginate: Salt of alginic acid, an edible sugar found in brown algae. Alginate is usually combined with sodium (i.e., sodium alginate).

alveolar sac: *See* **alveolus.**

alveolus: One of the millions of hollow, distensible cup-shaped cavities in the lungs where pulmonary gas exchange occurs.

aquifer: Body of underground rock, gravel, or soil whose pores are entirely saturated (i.e., connected) with groundwater, allowing the water to flow through the rock matrix and release the water (e.g., when pumped through a well).

arborescent: In a plant, trait or quality of being shaped like a tree or looking like a tree. Example: an arborescent fern.

atomizer: A device that produces a spray of droplets.

ballpoint pen: A pen that dispenses ink (usually in paste form) over a metal ball at its point.

base pair: In the DNA structure, each pair of nucleotide bases (A-T or G-C) in the "ladder rung" of DNA.

basilar membrane: A membrane within the cochlea of the inner ear that separates the cochlear and tympanic ducts, moving up and down in response to incoming sound waves.

Bernoulli's principle: A principle empirically obtained by Daniel Bernoulli which states that an increase in the speed of a fluid coincides with a decrease in static pressure or the fluid's gravitational energy (i.e., elevation). Bernoulli published it in 1738, and Leonhard Euler derived its equation in 1752.

bioink: A 3D printable ink that contains biological materials, usually proteins, a biological hydrogel, and/or cells.

bioprinter: A 3D printer that combines cells and biological materials to fabricate live biological tissue, usually to imitate natural tissue characteristics.

Birmingham gauge: In medicine, a measurement system that specifies the outside diameter of hypodermic/injection needles, catheters, cannulae, and suture wires in units that increase with decreasing diameter. For example, a needle of gauge 25 has an outer diameter of 0.508 millimeters, and one of gauge 26 has an outer diameter of 0.457 millimeters.

birome: (In Argentina) ballpoint pen.

bubble: A small globule of gas surrounded by liquid.

Caenorhabditis elegans: A transparent nematode used as a model organism in cell biology and the first multicellular organism whose genome was sequenced.

carburetor: The component of a combustion engine in which fuel becomes mixed with air.

chemotherapy: a type of cancer treatment that uses intracellular poisons to inhibit cell division or induce DNA damage.

circulating tumor cell: A tumor cell that has sloughed off the primary tumor and extravasates into and circulates in the blood, potentially becoming a seed for the subsequent growth of additional tumors (i.e., metastases) in distant organs.

cleanroom: A dust-free laboratory, typically consisting of filtered-air systems and appropriate behavior protocols (e.g., wearing gloves, booties, a gown, and a hair cover) to prevent the introduction of small particles.

cloud manufacturing: A new manufacturing paradigm that integrates a distributed network of advanced manufacturing nodes as well as design and virtual testing capabilities to optimize manufacturing efficiency.

cochlea: In vertebrates, the part of the inner ear involved in hearing. In mammals, this cavity is shaped like a small tube coiled into a spiral resembling a snail shell (hence its name).

cochlear duct: Cavity inside the cochlea located between the tympanic and the vestibular duct.

computed tomography scan: *See* **CT scan.**

condensation: Process by which a gas substance converts to its liquid form.

continuous glucose monitor: A glucose meter continuously connected to a patch affixed to the skin via a microneedle.

CT scan: *Abbreviation for* computed tomography scan, a medical imaging technique used to obtain detailed internal 3D images of the body.

CTC: *See* **circulating tumor cell.**

cytosine: *See* **nucleotide base.**

diabetes: A disease characterized by the body's inability to adjust the glucose levels in the blood.

diabetes mellitus: *See* **diabetes.**

dialysate: In kidney dialysis, the liquid into which the blood impurities pass as they cross the membrane.

dialysis: *See* **hemodialysis.**

diffusion: The random motion of molecules in a gas or dissolved in a liquid that accounts for their tendency to spread far away from their initial position and toward regions of lower concentration as they bounce with each other.

digital manufacturing: An integrated approach to manufacturing centered around a computer system that coordinates and/or automates the design, modeling, and fabrication of the part, which is encoded as a digital file from the beginning of the design phase.

direct ink writing: Extrusion-based 3D printing technique in which a filament of a paste-like material is extruded at room temperature from a small nozzle; the nozzle is moved across a platform to build the object layer by layer.

distributed manufacturing: Any form of decentralized manufacturing using a network of geographically dispersed manufacturing facilities, usually coordinated with information technology.

diya: An oil lamp made from clay or mud with a cotton wick dipped in oil or butter.

DNA polymerase: An enzyme that catalyzes DNA synthesis from deoxyribonucleotides, the DNA building blocks.

DNA sequencing: Any method or technology used to determine the order of the four bases (adenine, guanine, cytosine, and thymine) in DNA.

Drop-Seq: Abbreviation for "droplet sequencing," a method based on capturing single cells and sets of uniquely barcoded primer beads together in tiny droplets, enabling large-scale, highly parallel analysis of which genes are activated or repressed in every cell.

electrochemical reaction: A chemical reaction driven by an electrical potential difference, as in electrolysis, or that results in a potential difference, as in an electric battery.

electrolyte: In physiology, the ionized or ionizable constituents of a living cell, blood, or other organic matter.

electronic fuel injector: A type of fuel injector that is controlled electronically. The Bendix Electrojector, introduced in 1957, was the first electronic fuel injector. It used analog electronics for the control system. The Bosch Motronic injection system introduced in 1979 was the first mass-produced system to use digital electronics.

emphysema: In pulmonology, the enlargement of air spaces (alveoli) in the lungs.

emulsion: A fluid system in which liquid droplets are dispersed in a different liquid in which they usually do not mix, such as oil droplets in water.

EpCAM: *Abbreviation for* epithelial cell adhesion molecule, a protein on the surface of epithelial cells that is used as a marker to identify cancer cells of epithelial origin.

etching: In microfabrication, a technique to chemically remove layers from the surface of a chip during manufacturing.

eunuch: A person who has been castrated, often with a specific social motive.

fascicle: A complex microneedle-like organ used by female mosquitoes to pierce the skin of animals and extract blood from them.

fermentation: A metabolic process that produces chemical changes in organic substances through the action of enzymes. A well-known example is ethanol fermentation, which is used to produce alcoholic beverages and to make bread dough rise.

flow cytometer: Apparatus used to perform flow cytometry.

flow cytometry: A technique based on flowing single cells or particles through a glass capillary at high throughput to detect and measure the physical and chemical characteristics of those cells or particles.

fountain pen: A writing instrument that uses a metal nib to apply water-based ink to paper from an internal reservoir.

fuel injector: An internal combustion engine component that injects fuel into the engine at high speed. All diesel engines incorporate fuel injectors.

fused deposition modeling: *See* **fused filament fabrication.**

fused filament fabrication: A 3D printing process in which a filament of thermoplastic material is continuously heat-extruded from a small nozzle while the nozzle is moved across a platform to build an object layer by layer.

genome: The ensemble of all the genetic information of an organism.

gill: Respiratory organ that many aquatic organisms use to extract dissolved oxygen from water and to excrete carbon dioxide.

gland: A group of cells in an animal's body that synthesizes substances (e.g., hormones) for release into the bloodstream (endocrine gland) or into cavities inside the body or its outer surface (exocrine gland).

glomerulus: Basic filtration unit of the kidney. A capsule surrounds each glomerulus for urine collection.

glucometer: A glucose meter—that is, a device that measures glucose levels in the blood.

glucose oxidase: Enzyme that catalyzes glucose oxidation—that is, the reaction of glucose with oxygen.

guanine: *See* **nucleotide base.**

hCG: *Abbreviation for* human chorionic gonadotropin.

hemodialysis: A process for purifying the blood of a person whose kidneys are not working normally.

hertz: Unit of frequency. One hertz (abbreviated Hz) equals one cycle or vibration per second.

hollow fiber: A small, semipermeable capillary membrane. The membrane wall is typically permeable to low-molecular-weight ions and impermeable to molecules with a molecular weight larger than 10–30 kDa. These hollow fiber membranes are often bundled in a parallel array and housed within tubular polycarbonate shells to create hollow fiber bioreactor cartridges.

hormone: A class of signaling molecules in multicellular organisms that are sent to distant organs by complex biological processes to regulate physiology and behavior.

human chorionic gonadotropin: A hormone present in the blood and urine of pregnant women. This hormone is used to detect whether a woman is pregnant.

hydrogel: A 3D network of hydrophilic polymers that can swell in water and hold a large amount of water while maintaining the structure due to chemical or physical cross-linking of individual polymer chains.

hydrophilic: Tending to mix with water or be wetted by water.

hydrophilicity: The property of a hydrophilic material or its surface.

hydrophobic: Tending to repel water or fail to mix with water.

hydrophobicity: The property of a hydrophobic material or its surface.

hypodermic needle: Needle intended to penetrate the skin (subcutaneous use).

hypoglycemic attack: The condition caused by having too little glucose in the blood.

immunoassay: A biological measurement that is made with the help of antibodies. For example, "hCG immunoassay" means a measurement that detects the hormone hCG employing antibodies against hCG.

immunotherapy: The activation or suppression of the immune system to treat a disease such as cancer or an autoimmune disease.

inner ear: The innermost part of the vertebrate ear, mainly responsible for detecting sound, motion, and balance.

insulin: A hormone that tells your cells to absorb glucose from the blood.

interstitial fluid: The body fluid between blood vessels and cells, containing nutrients from blood capillaries by diffusion and holding waste products discharged by cells due to metabolism. Interstitial fluid has a different composition in different tissues and different areas of the body.

karst: In geology, a topography formed from the dissolution of soluble carbonate rocks such as limestone, dolomite, or gypsum. Many underground caves are examples of karst formations.

latex: A milky fluid that certain plants produce to defend themselves against herbivorous insects.

lung-on-a-chip: An organ-on-a-chip that simulates one or more of the lung functions.

lungs: The primary organs of the respiratory system in most animals. They extract oxygen from the air and transfer it into the bloodstream. They also release carbon dioxide from the bloodstream into the atmosphere. This dual process of gas exchange is diffusion-based.

lymphatic system: A system composed of lymph vessels (channels) and intervening lymph nodes whose function, like the venous system, is to return interstitial fluid from the tissues to be recirculated. At the origin of the fluid-return process, interstitial fluid enters the lymph capillaries.

Mandarin: An official in any of the nine top grades of the former imperial Chinese civil service. The name means "bossy" in Portuguese. The Portuguese gave this name to these officials because of the great zeal with which they performed their jobs.

MEMS: *See* **microelectromechanical systems.**

metastasis: Process by which cancer cells spread from a primary tumor to distant sites in the body, resulting in the development of secondary tumors.

metered-dose inhaler: A type of atomizer invented in 1955 by George Maison, the president of Riker Laboratories, to facilitate the dispensing of precise quantities of asthma medicine.

microelectromechanical systems: Microscopic or miniature devices incorporating electronic and/or mechanical parts, including devices that have fluidic, biological, and/or chemical functionalities.

microfabrication: The process of fabricating miniature structures of micrometer scales and smaller—for example, for microelectronics and microfluidics.

microneedles: Microfabricated needles.

Navier–Stokes equations: Equations that describe the motion of viscous fluids.

nebulizer: *See* **atomizer.**

nematode: Roundworms that inhabit a broad range of environments.

next-generation sequencing: A DNA sequencing method based on fragmenting the DNA into millions of parts and sequencing all the parts simultaneously by sequentially synthesizing, nucleotide by nucleotide, each DNA fragment. Because each nucleotide base is labeled with a different fluorophore color, the four DNA synthesis reactions corresponding to the four nucleotide bases (adenine, cytosine, thymine, and guanine) being added at each step can be monitored by fluorescence microscopy in an automated microfluidic setup.

non-small-cell lung cancer: A class of lung cancer types that accounts for approximately 85% of all lung cancers and that is relatively insensitive to chemotherapy.

nucleotide: The monomer unit of nucleic acids such as DNA. It comprises three distinctive chemical subunits: a five-carbon sugar molecule, a nucleotide base, and one phosphate group. The sugar molecule and the phosphate group form the double helix backbone, and the nucleotide bases form pairs inside the helix, encoding the genetic information.

nucleotide base: One of the three components of the nucleotides forming DNA. In DNA, the base can be adenine (A), cytosine (C), guanine (G), or thymine (T). Adenine always pairs with thymine, and guanine always pairs with cytosine.

organ of Corti: A tissue structure in the cochlea of the inner ear that produces nerve impulses in response to sound vibrations.

organ-on-a-chip: A microfluidic device that operates as a living avatar of a given human organ, typically designed to simulate or replace one or more organ functions outside the human body.

oval window: One of the two openings from the middle ear into the inner ear, allowing fluid in the cochlea to move, ensuring that hair cells of the basilar membrane will be stimulated and that audition will occur.

oxidation: The loss of electrons in a chemical reaction.

paper microfluidics: Microfluidic technology in which fluid flows within a paper substrate, typically powered by the wicking action of the paper.

parchment: Writing material made from specially prepared untanned skins of animals—primarily sheep, calves, and goats. The word derives from the ancient Greek city of Pergamon (present-day Bergama in Turkey).

PCR: *Abbreviation for* polymerase chain reaction.

percolation: Movement of a fluid through a porous material.

phloem: In land plants, the vascular system that transports the sugar sucrose and other soluble organic compounds made during photosynthesis to the rest of the plant.

photolithography: In microelectronics and microfluidics fabrication, photolithography designates a set of techniques that use light to produce small patterns of suitable materials over a flat substrate. During photolithography, ultraviolet light transfers a geometric design through an optical mask (the *photomask*) to a light-sensitive chemical (the *photoresist*) coated on the substrate. The goal of photolithography is generally to protect selected areas of the substrate during subsequent processing, such as etching or metallization.

photomask: A transparent plate with opaque areas used in photolithography that allows ultraviolet light to shine through in a defined pattern.

photoresist: A light-sensitive material used in photolithography to form a patterned coating on a surface.

photosynthesis: The process by which green plants and some other organisms use sunlight, water, and carbon dioxide to create oxygen and simple sugars (e.g., glucose and fructose). These simple sugars are combined into more complex sugars, such as sucrose, for transport and storage.

pneumonia: Inflammatory condition of the lung primarily affecting the alveoli. Symptoms typically include some combination of cough, shortness of breath, chest pain, and fever.

polymerase chain reaction: A method widely used to rapidly make millions to billions of copies of a specific DNA sample, allowing scientists to exponentially amplify a very small DNA sample (or a part of it) sufficiently to enable its detailed study.

primary cancer/primary tumor: A tumor growing at the anatomical site where the tumor first originated.

primer: A nucleic acid strand (or related molecule) that serves as a starting point for replication in DNA synthesis.

prognosis: In medicine, a prediction of how a disease is likely to affect the patient, especially their recovery.

qalam: A reed pen.

quill: A writing tool made from a feather, preferably a wing feather of a large bird.

Rayleigh instability: Fluid phenomenon responsible for the breakup of a fluid jet or stream into droplets due to the minimization of surface area by the surface tension of the liquid.

redox reaction: A chemical reaction between two chemical species that react by exchanging one or more electrons. The species that loses electrons is said to be oxidized, whereas the species that gains electrons is said to be reduced.

reduction: The gain of electrons in a chemical reaction.

reed: A tall, slender-leaved plant of the grass family that grows in water or on marshy ground.

refraction: A phenomenon that occurs at the interface between two different transparent materials contacting each other when illuminated by light and consists of light "bending" (i.e., a change of direction in the light's traveling direction).

sap: Fluid transported in the xylem or phloem vessels of a plant. Xylem sap contains water, inorganic ions, and a few organic chemicals, whereas phloem sap contains water and sugars.

secondary cancer/secondary tumor: A tumor that, as a result of metastasis, is growing at an anatomical site that is distant from the primary tumor site.

sky lantern: A small hot air balloon made of paper, with an opening at the bottom where a small fire is suspended.

spermaceti: A waxy liquid obtained from the spermaceti organ of sperm whales.

spermaceti oil: *See* **spermaceti.**

spermaceti organ: An organ present in the heads of sperm whales that contains a waxy liquid called spermaceti.

spherification: A chemical process used in gastronomy and biotechnology that consists of the gelation of sodium alginate by adding a salt such as calcium chloride.

spheroid: In tissue culture, a small spherical mass of live cells used as a miniature 3D tissue model.

spray-nozzle carburetor: A type of carburetor invented in 1893 by Wilhelm Maybach that draws fuel into the intake airstream by using the suction the intake air creates as it accelerates through a constriction (a Venturi). The collision of the drawn fuel jet with the airstream produces a spray of fuel droplets.

stage of a cancer: A classification that characterizes the extent to which a tumor has grown and spread. The classification goes from Stage I (localized cancer) to Stage IV (metastasized tumor).

surface runoff: In hydrology, the flow of excess rainwater not infiltrating the ground over its surface.

surface tension: The tendency of a liquid surface at rest to shrink into the minimum surface area possible.

tallow: Fat from ruminants, including cattle, bison, and lamb.

thymine: *See* **nucleotide base.**

transpiration: The biological process by which water in plants evaporates as water vapor from the surface of leaves.

tympanic duct: Cavity in the inner ear of vertebrates that connects the end of the vestibular duct with the round window. This fluid-filled duct and the flexible round window allow the eardrum to vibrate.

type 1 diabetes: Type of diabetes caused by the pancreas' inability to make enough insulin.

type 2 diabetes: Type of diabetes caused by the body's inability to respond normally to the insulin made by the pancreas.

ultrasound: High-frequency sound that is inaudible to humans.

vascular plants: Land plants that have a set of specialized vascular tissues to conduct water and minerals throughout the plant (the *xylem*) and also a separate set of specialized tissues (the *phloem*) to distribute the products of photosynthesis.

Venturi effect: The reduction in fluid pressure that results when fluid flows through a constricted section of a pipe. Due to Bernoulli's principle, there is a simultaneous increase in the fluid velocity at the constriction.

vestibular duct: Cavity inside the cochlea of the inner ear that conducts sound vibrations to the cochlear duct.

viscosity: A measure of the resistance to deformation of a fluid at a given rate.

water lantern: A lamp designed to float on the water's surface.

wearable: Technology (e.g., microelectronic or microfluidic device) designed to function or be used while worn.

wick: The cord in a candle or oil lamp to light the flame.

xylem: In land plants, the vascular system that transports water and soluble mineral nutrients from the roots throughout the plant.

Index

For the benefit of digital users, indexed terms that span two pages (e.g., 52–53) may, on occasion, appear on only one of those pages.

204